T0250297

R Markdown
The Definitive Guide

Chapman & Hall/CRC
The R Series

Series Editors

John M. Chambers, Department of Statistics Stanford University Stanford, California, USA

Torsten Hothorn, Division of Biostatistics University of Zurich Switzerland

Duncan Temple Lang, Department of Statistics University of California, Davis, California, USA

Hadley Wickham, RStudio, Boston, Massachusetts, USA

Recently Published Titles

bookdown: Authoring Books and Technical Documents with R Markdown
Yihui Xie

Testing R Code
Richard Cotton

R Primer, Second Edition
Claus Thorn Ekstrøm

Flexible Regression and Smoothing: Using GAMLSS in R
Mikis D. Stasinopoulos, Robert A. Rigby, Gillian Z. Heller, Vlasios Voudouris, Fernanda De Bastiani

The Essentials of Data Science: Knowledge Discovery Using R
Graham J. Williams

blogdown: Creating Websites with R Markdown
Yihui Xie, Alison Presmanes Hill, Amber Thomas

Handbook of Educational Measurement and Psychometrics Using R
Christopher D. Desjardins, Okan Bulut

Displaying Time Series, Spatial, and Space-Time Data with R, Second Edition
Oscar Perpinan Lamigueiro

Reproducible Finance with R
Jonathan K. Regenstein, Jr

R Markdown: The Definitive Guide
Yihui Xie, J.J. Allaire, Garrett Grolemund

For more information about this series, please visit:
https://www.crcpress.com/go/the-r-series

R Markdown
The Definitive Guide

Yihui Xie
J. J. Allaire
Garrett Grolemund

CRC Press
Taylor & Francis Group
Boca Raton London New York

CRC Press is an imprint of the
Taylor & Francis Group, an **informa** business

A CHAPMAN & HALL BOOK

To Jung Jae-sung (1982 – 2018),

a remarkably hard-working badminton player with a remarkably simple playing style

Contents

List of Tables

List of Figures

Preface

The document format "R Markdown" was first introduced in the **knitr** package (Xie, 2015, 2018d) in early 2012. The idea was to embed code chunks (of R or other languages) in Markdown documents. In fact, **knitr** supported several authoring languages from the beginning in addition to Markdown, including LaTeX, HTML, AsciiDoc, reStructuredText, and Textile. Looking back over the five years, it seems to be fair to say that Markdown has become the most popular document format, which is what we expected. The simplicity of Markdown clearly stands out among these document formats.

However, the original version of Markdown invented by John Gruber[1] was often found overly simple and not suitable to write highly technical documents. For example, there was no syntax for tables, footnotes, math expressions, or citations. Fortunately, John MacFarlane created a wonderful package named Pandoc (http://pandoc.org) to convert Markdown documents (and many other types of documents) to a large variety of output formats. More importantly, the Markdown syntax was significantly enriched. Now we can write more types of elements with Markdown while still enjoying its simplicity.

In a nutshell, R Markdown stands on the shoulders of **knitr** and Pandoc. The former executes the computer code embedded in Markdown, and converts R Markdown to Markdown. The latter renders Markdown to the output format you want (such as PDF, HTML, Word, and so on).

The **rmarkdown** package (Allaire et al., 2018c) was first created in early 2014. During the past four years, it has steadily evolved into a relatively complete ecosystem for authoring documents, so it is a good time for us to provide a definitive guide to this ecosystem now. At this point, there are a large number of tasks that you could do with R Markdown:

- Compile a single R Markdown document to a report in different formats, such as PDF, HTML, or Word.

---

[1]https://en.wikipedia.org/wiki/Markdown

- Create notebooks in which you can directly run code chunks interactively.

- Make slides for presentations (HTML5, LaTeX Beamer, or PowerPoint).

- Produce dashboards with flexible, interactive, and attractive layouts.

- Build interactive applications based on Shiny.

- Write journal articles.

- Author books of multiple chapters.

- Generate websites and blogs.

There is a fundamental assumption underneath R Markdown that users should be aware of: we assume it suffices that only a limited number of features are supported in Markdown. By "features", we mean the types of elements you can create with native Markdown. The limitation is a great feature, not a bug. R Markdown may not be the right format for you if you find these elements not enough for your writing: paragraphs, (section) headers, block quotations, code blocks, (numbered and unnumbered) lists, horizontal rules, tables, inline formatting (emphasis, strikeout, superscripts, subscripts, verbatim, and small caps text), LaTeX math expressions, equations, links, images, footnotes, citations, theorems, proofs, and examples. We believe this list of elements suffice for most technical and non-technical documents. It may not be impossible to support other types of elements in R Markdown, but you may start to lose the simplicity of Markdown if you wish to go that far.

Epictetus once said, *"Wealth consists not in having great possessions, but in having few wants."* The spirit is also reflected in Markdown. If you can control your preoccupation with pursuing typesetting features, you should be much more efficient in writing the content and can become a prolific author. It is entirely possible to succeed with simplicity. Jung Jae-sung was a legendary badminton player with a remarkably simply playing style: he did not look like a talented player and was very short compared to other players, so most of the time you would just see him jump three feet off the ground and smash like thunder over and over again in the back court until he beats his opponents.

Please do not underestimate the customizability of R Markdown because of the simplicity of its syntax. In particular, Pandoc templates can be surprisingly powerful, as long as you understand the underlying technologies such as LaTeX and CSS, and are willing to invest time in the appearance of your output documents (reports, books, presentations, and/or websites). As one example, you may check out the PDF report[2] of the 2017 Employer Health Benefits Survey[3]. It looks fairly sophisticated, but was actually produced via **bookdown** (Xie, 2016), which is an R Markdown extension. A custom LaTeX template and a lot of LaTeX tricks were used to generate this report. Not surprisingly, this very book that you are reading right now was also written in R Markdown, and its full source is publicly available in the GitHub repository `https://github.com/rstudio/rmarkdown-book`.

R Markdown documents are often portable in the sense that they can be compiled to multiple types of output formats. Again, this is mainly due to the simplified syntax of the authoring language, Markdown. The simpler the elements in your document are, the more likely that the document can be converted to different formats. Similarly, if you heavily tailor R Markdown to a specific output format (e.g., LaTeX), you are likely to lose the portability, because not all features in one format work in another format.

Last but not least, your computing results will be more likely to be reproducible if you use R Markdown (or other **knitr**-based source documents), compared to the manual cut-and-paste approach. This is because the results are dynamically generated from computer source code. If anything goes wrong or needs to be updated, you can simply fix or update the source code, compile the document again, and the results will automatically updated. You can enjoy reproducibility and convenience at the same time.

How to read this book

This book may serve you better as a reference book than a textbook. It contains a large number of technical details, and we do not expect you to read it from beginning to end, since you may easily feel overwhelmed. Instead, think about your background and what you want to do first, and go to the relevant chapters or sections. For example:

- I just want to finish my course homework (Chapter 2 should be more than enough for you).

---

[2]`http://files.kff.org/attachment/Report-Employer-Health-Benefits-Annual-Survey-2017`

[3]`https://www.kff.org/health-costs/report/2017-employer-health-benefits-survey/`

- I know this is an R Markdown book, but I use Python more than R (Go to Section 2.7.1).

- I want to embed interactive plots in my reports, or want my readers to be able change my model parameters interactively and see results on the fly (Check out Section 2.8).

- I know the output format I want to use, and I want to customize its appearance (Check out the documentation of the specific output format in Chapter 3 or Chapter 4). For example, I want to customize the template for my PowerPoint presentation (Go to Section 4.4.1).

- I want to build a business dashboard highlighting some key figures and indicators (Go to Chapter 5).

- I heard about yolo = TRUE from a friend, and I'm curious what that means in the **xaringan** package (Go to Chapter 7).

- I want to build a personal website (Go to Chapter 10), or write a book (Go to Chapter 12).

- I want to write a paper and submit to the Journal of Statistical Software (Go to Chapter 13).

- I want to build an interactive tutorial with exercises for my students to learn a topic (Go to Chapter 14).

- I'm familiar with R Markdown now, and I want to generate personalized reports for all my customers using the same R Markdown template (Try parameterized reports in Chapter 15).

- I know some JavaScript, and want to build an interface in R to call an interested JavaScript library from R (Learn how to develop HTML widgets in Chapter 16).

- I want to build future reports with a company branded template that shows our logo and uses our unique color theme (Go to Chapter 17).

If you are not familiar with R Markdown, we recommend that you read at least Chapter 2 to learn the basics. All the rest of the chapters in this book can be read in any order you desire. They are pretty much orthogonal to each other. However, to become familiar with R Markdown output formats, you may want to thumb through the HTML document format in Section 3.1, because many other formats share the same options as this format.

Structure of the book

This book consists of four parts. Part I covers the basics: Chapter 1 introduces how to install the relevant packages, and Chapter 2 is an overview of R Markdown, including the possible output formats, the Markdown syntax, the R code chunk syntax, and how to use other languages in R Markdown.

Part II is the detailed documentation of built-in output formats in the **rmarkdown** package, including document formats and presentation formats.

Part III lists about ten R Markdown extensions that enable you to build different applications or generate output documents with different styles. Chapter 5 introduces the basics of building flexible dashboards with the R package **flexdashboard**. Chapter 6 documents the **tufte** package, which provides a unique document style used by Edward Tufte. Chapter 7 introduces the **xaringan** package for another highly flexible and customizable HTML5 presentation format based on the JavaScript library remark.js. Chapter 8 documents the **revealjs** package, which provides yet another appealing HTML5 presentation format based on the JavaScript library reveal.js. Chapter 9 introduces a few output formats created by the R community, such as the **prettydoc** package, which features lightweight HTML document formats. Chapter 10 teaches you how to build websites using either the **blogdown** package or **rmarkdown**'s built-in site generator. Chapter 11 explains the basics of the **pkgdown** package, which can be used to quickly build documentation websites for R packages. Chapter 12 introduces how to write and publish books with the **bookdown** package. Chapter 13 is an overview of the **rticles** package for authoring journal articles. Chapter 14 introduces how to build interactive tutorials with exercises and/or quiz questions.

Part IV covers other topics about R Markdown, and some of them are advanced (in particular, Chapter 16). Chapter 15 introduces how to generate different reports with the same R Markdown source document and different parameters. Chapter 16 teaches developers how to build their own HTML widgets for interactive visualization and applications with JavaScript libraries. Chapter 17 shows how to create custom R Markdown and Pandoc templates so that you can fully customize the appearance and style of your output document. Chapter 18 explains how to create your own output formats if the existing formats do not meet your need. Chapter 19 shows how to combine the Shiny framework with R Markdown, so that your readers can interact with the reports by changing the values of certain input widgets and seeing updated results immediately.

Note that this book is intended to be a guide instead of the comprehensive documentation of all topics related to R Markdown. Some chapters are only overviews, and you may need to consult the full documentation elsewhere (often freely available online). Such examples include Chapters 5, 10, 11, 12, and 14.

Software information and conventions

The R session information when compiling this book is shown below:

```
xfun::session_info(c(
  'blogdown', 'bookdown', 'knitr', 'rmarkdown', 'htmltools',
  'reticulate', 'rticles', 'flexdashboard', 'learnr', 'shiny',
  'revealjs', 'pkgdown', 'tinytex', 'xaringan', 'tufte'
), dependencies = FALSE)
```

```
## R version 3.5.0 (2018-04-23)
## Platform: x86_64-apple-darwin15.6.0 (64-bit)
## Running under: macOS High Sierra 10.13.5
##
## Locale: en_US.UTF-8 / en_US.UTF-8 / en_US.UTF-8 / C / en_US.UTF-
8 / en_US.UTF-8
##
## Package version:
##   blogdown_0.6.12      bookdown_0.7.11
##   flexdashboard_0.5.1 htmltools_0.3.6
##   knitr_1.20.5         learnr_0.9.2
##   pkgdown_1.1.0        reticulate_1.8
##   revealjs_0.9         rmarkdown_1.10.2
##   rticles_0.4.2.9000   shiny_1.1.0
##   tinytex_0.5.8        tufte_0.3
##   xaringan_0.6.7
##
## Pandoc version: 2.2.1
```

We do not add prompts (> and +) to R source code in this book, and we comment out the text output with two hashes ## by default, as you can see from the R session information above. This is for your convenience when you want to copy and run the code (the text output will be ignored since it is commented out). Package names are in bold text (e.g., **rmarkdown**), and inline code and filenames are formatted in a typewriter font (e.g., `knitr::knit('foo.Rmd')`). Function names are followed by parentheses (e.g., `blogdown::serve_site()`). The double-colon operator `::` means accessing an object from a package.

"Rmd" is the filename extension of R Markdown files, and also an abbreviation of R Markdown in this book.

Acknowledgments

I started writing this book after I came back from the 2018 RStudio Conference in early February, and finished the first draft in early May. This may sound fast for a 300-page book. The main reason I was able to finish it quickly was that I worked full-time on this book for three months. My employer, RStudio, has always respected my personal interests and allowed me to focus on projects that I choose by myself. More importantly, I have been taught several lessons on how to become a professional software engineer since I joined RStudio as a fresh PhD, although the initial journey turned out to be painful.[4] It is a great blessing for me to work in this company.

The other reason for my speed was that JJ and Garrett had already prepared a lot of materials that I could adapt for this book. They had also been offering suggestions as I worked on the manuscript. In addition, Michael Harper[5] contributed the initial drafts of Chapters 12, 13, 15, 17, and 18. I would definitely not be able to finish this book so quickly without their help.

The most challenging thing to do when writing a book is to find large blocks of uninterrupted time. This is just so hard. Both others and myself could interrupt me. I do not consider my willpower to be strong: I read random articles, click on the endless links on Wikipedia, look at random Twitter mes-

---

[4] `https://yihui.name/en/2018/02/career-crisis/`
[5] `http://mikeyharper.uk`

sages, watch people fight on meaningless topics online, reply to emails all the time as if I were able to reach "Inbox Zero", and write random blog posts from time to time. The two most important people in terms of helping keep me on track are Tareef Kawaf (President of RStudio), to whom I report my progress on the weekly basis, and Xu Qin[6], from whom I really learned[7] the importance of making plans on a daily basis (although I still fail to do so sometimes). For interruptions from other people, it is impossible to isolate myself from the outside world, so I'd like to thank those who did not email me or ask me questions in the past few months and used public channels instead as I suggested[8]. I also thank those who did not get mad at me when my responses were extremely slow or even none. I appreciate all your understanding and patience. Besides, several users have started helping me answer GitHub and Stack Overflow questions related to R packages that I maintain, which is even better! These users include Marcel Schilling[9], Xianying Tan[10], Christophe Dervieux[11], and Garrick Aden-Buie[12], just to name a few. As someone who works from home, apparently I would not even have ten minutes of uninterrupted time if I do not send the little ones to daycare, so I want to thank all teachers at Small Miracles for freeing my daytime.

There have been a large number of contributors to the R Markdown ecosystem. More than 60 people[13] have contributed to the core package, **rmarkdown**. Several authors have created their own R Markdown extensions, as introduced in Part III of this book. Contributing ideas is no less helpful than contributing code. We have gotten numerous inspirations and ideas from the R community via various channels (GitHub issues, Stack Overflow questions, and private conversations, etc.). As a small example, Jared Lander, author of the book *R for Everyone*, does not meet me often, but every time he chats with me, I will get something valuable to work on. "How about writing books with R Markdown?" he asked me at the 2014 Strata conference in New York. Then we invented **bookdown** in 2016. "I really need fullscreen background images in ioslides. Look, Yihui, here are my ugly JavaScript hacks,[14]" he showed me

---

[6]http://home.uchicago.edu/~xuqin/

[7]https://d.cosx.org/d/419325

[8]https://yihui.name/en/2017/08/so-gh-email/

[9]https://yihui.name/en/2018/01/thanks-marcel-schilling/

[10]https://shrektan.com

[11]https://github.com/cderv

[12]https://www.garrickadenbuie.com

[13]https://github.com/rstudio/rmarkdown/graphs/contributors

[14]https://www.jaredlander.com/2017/07/fullscreen-background-images-in-ioslides-presentations/

on the shuttle to dinner at the 2017 RStudio Conference. A year later, background images were officially supported in ioslides presentations.

As I mentioned previously, R Markdown is standing on the shoulders of the giant, Pandoc. I'm always amazed by how fast John MacFarlane, the main author of Pandoc, responds to my GitHub issues. It is hard to imagine a person dealing with 5000 GitHub issues[15] over the years while maintaining the excellent open-source package and driving the Markdown standards forward. We should all be grateful to John and contributors of Pandoc.

As I was working on the draft of this book, I received a lot of helpful reviews from these reviewers: John Gillett (University of Wisconsin), Rose Hartman (UnderstandingData), Amelia McNamara (Smith College), Ariel Muldoon (Oregon State University), Yixuan Qiu (Purdue University), Benjamin Soltoff (University of Chicago), David Whitney (University of Washington), and Jon Katz (independent data analyst). Tareef Kawaf (RStudio) also volunteered to read the manuscript and provided many helpful comments. Aaron Simumba[16], Peter Baumgartner[17], and Daijiang Li[18] volunteered to carefully correct many of my typos. In particular, Aaron has been such a big helper with my writing (not limited to only this book) and sometimes I have to compete with him[19] in correcting my typos!

There are many colleagues at RStudio whom I want to thank for making it so convenient and even enjoyable to author R Markdown documents, especially the RStudio IDE team including J.J. Allaire, Kevin Ushey, Jonathan McPherson, and many others.

Personally I often feel motivated by members of the R community. My own willpower is weak, but I can gain a lot of power from this amazing community. Overall the community is very encouraging, and sometimes even fun, which makes me enjoy my job. For example, I do not think you can often use the picture of a professor for fun in your software, but the "desiccated baseR-er"[20] Karl Broman is an exception (see Section 7.3.6), as he allowed me to use a mysteriously happy picture of him.

Lastly, I want to thank my editor, John Kimmel, for his continued help with my fourth book. I think I have said enough about him and his team at Chap-

---

[15]https://github.com/jgm/pandoc
[16]https://asimumba.rbind.io
[17]http://peter.baumgartner.name
[18]https://daijiang.name
[19]https://github.com/rbind/yihui/commit/d8f39f7aa
[20]https://twitter.com/kwbroman/status/922545181634768897

man & Hall in my previous books. The publishing experience has always been so smooth. I just wonder if it would be possible someday that our meticulous copy-editor, Suzanne Lassandro, would fail to identify more than 30 issues for me to correct in my first draft. Probably not. Let's see.

Yihui Xie
Elkhorn, Nebraska

About the Authors

This book is primarily put together by me (Yihui Xie), making use of the existing R documentation of the **rmarkdown** package and the **rmarkdown** website, which were mainly contributed by J.J. Allaire and Garrett Grolemund.

Yihui Xie

Yihui Xie (`https://yihui.name`) is a software engineer at RStudio (`https://www.rstudio.com`). He earned his PhD from the Department of Statistics, Iowa State University. He is interested in interactive statistical graphics and statistical computing. As an active R user, he has authored several R packages, such as **knitr**, **bookdown**, **blogdown**, **xaringan**, **tinytex**, **animation**, **DT**, **tufte**, **formatR**, **fun**, **xfun**, **mime**, **highr**, **servr**, and **Rd2roxygen**, among which the **animation** package won the 2009 John M. Chambers Statistical Software Award (ASA). He also co-authored a few other R packages, including **shiny**, **rmarkdown**, and **leaflet**.

He has authored two books, *Dynamic Documents with knitr* (Xie, 2015), and *bookdown: Authoring Books and Technical Documents with R Markdown* (Xie, 2016), and co-authored the book, *blogdown: Creating Websites with R Markdown* (Xie et al., 2017).

In 2006, he founded the Capital of Statistics (`https://cosx.org`), which has grown into a large online community on statistics in China. He initiated the Chinese R conference in 2008, and has been involved in organizing R conferences in China since then. During his PhD training at Iowa State University, he won the Vince Sposito Statistical Computing Award (2011) and the Snedecor Award (2012) in the Department of Statistics.

He occasionally rants on Twitter (`https://twitter.com/xieyihui`), and most of the time you can find him on GitHub (`https://github.com/yihui`).

He enjoys spicy food as much as classical Chinese literature.

J.J. Allaire

J.J. Allaire is the founder of RStudio and the creator of the RStudio IDE. J.J. is an author of several packages in the R Markdown ecosystem including **rmarkdown**, **flexdashboard**, **learnr**, and **radix**.

Garrett Grolemund

Garrett Grolemund is the co-author of *R for Data Science* and author of *Hands-On Programming with R*. He wrote the **lubridate** R package and works for RStudio as an advocate who trains engineers to do data science with R and the Tidyverse. If you use R yourself, you may recognize Garrett from his video courses on Datacamp.com and O'Reilly media, or for his series of popular R cheatsheets distributed by RStudio.

Garrett earned his PhD in Statistics from Rice University in 2012 under the guidance of Hadley Wickham. Before that, he earned a Bachelor's degree in Psychology from Harvard University and briefly attended law school. Garrett has been one of the foremost promoters of Shiny, R Markdown, and the Tidyverse, documenting and explaining each in detail.

Part I

Get Started

1

Installation

We assume you have already installed R (`https://www.r-project.org`) (R Core Team, 2018) and the RStudio IDE (`https://www.rstudio.com`). RStudio is not required but recommended, because it makes it easier for an average user to work with R Markdown. If you do not have RStudio IDE installed, you will have to install Pandoc (`http://pandoc.org`), otherwise there is no need to install Pandoc separately because RStudio has bundled it. Next you can install the **rmarkdown** package in R:

```r
# Install from CRAN
install.packages('rmarkdown')

# Or if you want to test the development version,
# install from GitHub
if (!requireNamespace("devtools"))
  install.packages('devtools')
devtools::install_github('rstudio/rmarkdown')
```

If you want to generate PDF output, you will need to install LaTeX. For R Markdown users who have not installed LaTeX before, we recommend that you install TinyTeX (`https://yihui.name/tinytex/`):

```r
install.packages("tinytex")
tinytex::install_tinytex()  # install TinyTeX
```

TinyTeX is a lightweight, portable, cross-platform, and easy-to-maintain La-TeX distribution. The R companion package **tinytex** (Xie, 2018f) can help you automatically install missing LaTeX packages when compiling LaTeX or R Markdown documents to PDF, and also ensures a LaTeX document is compiled for the correct number of times to resolve all cross-references. If you do not understand what these two things mean, you should probably follow our recommendation to install TinyTeX, because these details are often not worth your time or attention.

With the **rmarkdown** package, RStudio/Pandoc, and LaTeX, you should be able to compile most R Markdown documents. In some cases, you may need other software packages, and we will mention them when necessary.

2

Basics

R Markdown provides an authoring framework for data science. You can use a single R Markdown file to both

- save and execute code, and

- generate high quality reports that can be shared with an audience.

R Markdown was designed for easier reproducibility, since both the computing code and narratives are in the same document, and results are automatically generated from the source code. R Markdown supports dozens of static and dynamic/interactive output formats.

If you prefer a video introduction to R Markdown, we recommend that you check out the website https://rmarkdown.rstudio.com, and watch the videos in the "Get Started" section, which cover the basics of R Markdown.

Below is a minimal R Markdown document, which should be a plain-text file, with the conventional extension .Rmd:

```
---
title: "Hello R Markdown"
author: "Awesome Me"
date: "2018-02-14"
output: html_document
---


This is a paragraph in an R Markdown document.

Below is a code chunk:

```{r}
fit = lm(dist ~ speed, data = cars)
b   = coef(fit)
```
```

```
plot(cars)
abline(fit)
```

```
The slope of the regression is `r b[1]`.
```

You can create such a text file with any editor (including but not limited to RStudio). If you use RStudio, you can create a new Rmd file from the menu `File -> New File -> R Markdown`.

There are three basic components of an R Markdown document: the metadata, text, and code. The metadata is written between the pair of three dashes `---`. The syntax for the metadata is YAML (YAML Ain't Markup Language, `https://en.wikipedia.org/wiki/YAML`), so sometimes it is also called the YAML metadata or the YAML frontmatter. Before it bites you hard, we want to warn you in advance that indentation matters in YAML, so do not forget to indent the sub-fields of a top field properly. See the Appendix B.2[1] of Xie (2016) for a few simple examples that show the YAML syntax.

The body of a document follows the metadata. The syntax for text (also known as prose or narratives) is Markdown, which is introduced in Section 2.5. There are two types of computer code, which are explained in detail in Section 2.6:

- A code chunk starts with three backticks like ```` ```{r} ```` where r indicates the language name,[2] and ends with three backticks. You can write chunk options in the curly braces (e.g., set the figure height to 5 inches: ```` ```{r, fig.height=5}````).

- An inline R code expression starts with `` `r `` and ends with a backtick `` ` ``.

Figure 2.1 shows the above example in the RStudio IDE. You can click the `Knit` button to compile the document (to an HTML page). Figure 2.2 shows the output in the RStudio Viewer.

Now please take a closer look at the example. Did you notice a problem? The object b is the vector of coefficients of length 2 from the linear regression; b[1] is actually the intercept, and b[2] is the slope! This minimal example shows you why R Markdown is great for reproducible research: it includes

---

[1] `https://bookdown.org/yihui/bookdown/r-markdown.html`

[2] It is not limited to the R language; see Section 2.7 for how to use other languages.

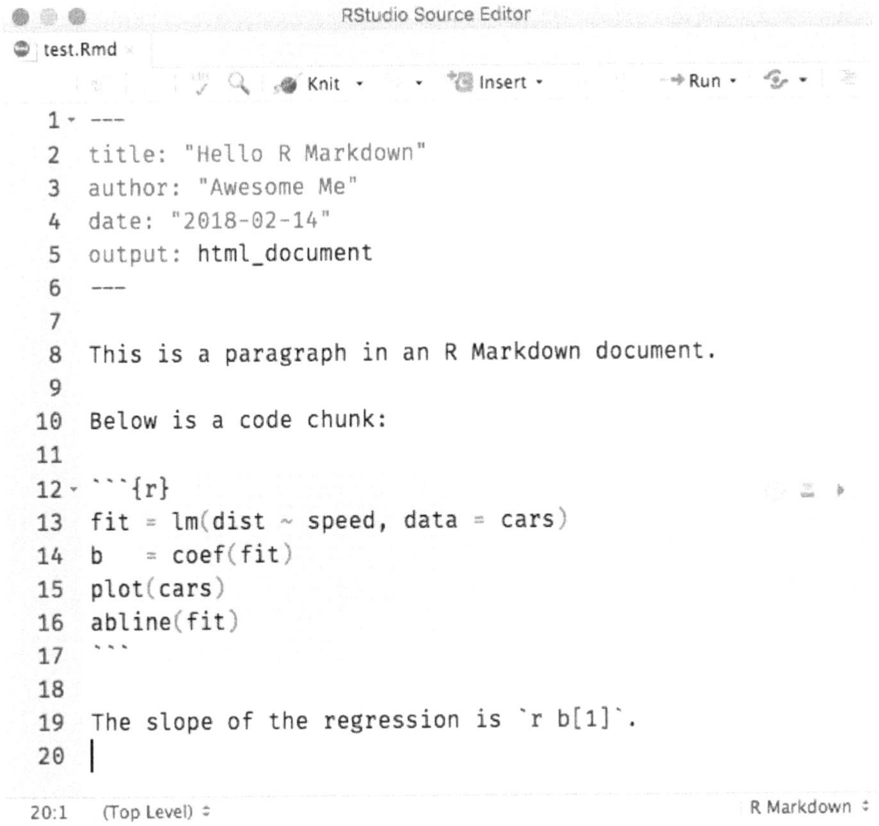

```
●　●　●                          RStudio Source Editor
⊜ test.Rmd

          ⌄⃗  Q   Knit ▾      ▾  ⁺⊡ Insert ▾           → Run ▾  ⌇ ▾    ≛
  1 ▾ ---
  2   title: "Hello R Markdown"
  3   author: "Awesome Me"
  4   date: "2018-02-14"
  5   output: html_document
  6   ---
  7
  8   This is a paragraph in an R Markdown document.
  9
 10   Below is a code chunk:
 11
 12 ▾ ```{r}                                              ⌇  ≛  ▸
 13   fit = lm(dist ~ speed, data = cars)
 14   b   = coef(fit)
 15   plot(cars)
 16   abline(fit)
 17   ```
 18
 19   The slope of the regression is `r b[1]`.
 20   |

  20:1    (Top Level) ⌖                                R Markdown ⌖
```

FIGURE 2.1: A minimal R Markdown example in RStudio.

the source code right inside the document, which makes it easy to discover and fix problems, as well as update the output document. All you have to do is change b[1] to b[2], and click the Knit button again. Had you copied a number -17.579 computed elsewhere into this document, it would be very difficult to realize the problem. In fact, I had used this example a few times by myself in my presentations before I discovered this problem during one of my talks, but I discovered it anyway.

Although the above is a toy example, it could become a horror story if it happens in scientific research that was not done in a reproducible way (e.g., cut-and-paste). Here are two of my personal favorite videos on this topic:

- "A reproducible workflow" by Ignasi Bartomeus and Francisco Rodríguez-Sánchez (https://youtu.be/s3JldKoA0zw). It is a 2-min

Files   Plots   Packages   Help   Viewer   Jobs

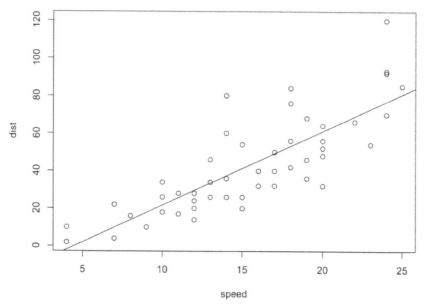

Hello R Markdown

Awesome Me

2018-02-14

This is a paragraph in an R Markdown document.

Below is a code chunk:

```
fit = lm(dist ~ speed, data = cars)
b = coef(fit)
plot(cars)
abline(fit)
```

The slope of the regression is -17.5790949.

FIGURE 2.2: The output document of the minimal R Markdown example in RStudio.

video that looks artistic but also shows very common and practical problems in data analysis.

- "The Importance of Reproducible Research in High-Throughput Biology" by Keith Baggerly (`https://youtu.be/7gYIs7uYbMo`). You will be impressed by both the content and the style of this lecture. Keith Baggerly and Kevin Coombes were the two notable heroes in revealing the Duke/Potti scandal[3], which was described as "one of the biggest medical research frauds ever" by the television program "60 Minutes".

It is fine for humans to err (in computing), as long as the source code is readily available.

2.1 Example applications

Now you have learned the very basic concepts of R Markdown. The idea should be simple enough: interweave narratives with code in a document, knit the document to dynamically generate results from the code, and you will get a report. This idea was not invented by R Markdown, but came from an early programming paradigm called "Literate Programming" (Knuth, 1984).

Due to the simplicity of Markdown and the powerful R language for data analysis, R Markdown has been widely used in many areas. Before we dive into the technical details, we want to show some examples to give you an idea of its possible applications.

2.1.1 Airbnb's knowledge repository

Airbnb uses R Markdown to document all their analyses in R, so they can combine code and data visualizations in a single report (Bion et al., 2018). Eventually all reports are carefully peer-reviewed and published to a company knowledge repository, so that anyone in the company can easily find analyses relevant to their team. Data scientists are also able to learn as much

---

[3]`https://en.wikipedia.org/wiki/Anil_Potti`

as they want from previous work or reuse the code written by previous authors, because the full R Markdown source is available in the repository.

2.1.2   Homework assignments on RPubs

A huge number of homework assignments have been published to the website `https://RPubs.com` (a free publishing platform provided by RStudio), which shows that R Markdown is easy and convenient enough for students to do their homework assignments (see Figure 2.3). When I was still a student, I did most of my homework assignments using Sweave, which was a much earlier implementation of literate programming based on the S language (later R) and LaTeX. I was aware of the importance of reproducible research but did not enjoy LaTeX, and few of my classmates wanted to use Sweave. Right after I graduated, R Markdown was born, and it has been great to see so many students do their homework in the reproducible manner.

In a 2016 JSM (Joint Statistical Meetings) talk, I proposed that course instructors could sometimes intentionally insert some wrong values in the source data before providing it to the students for them to analyze the data in the homework, then correct these values the next time, and ask them to do the analysis again. This way, students should be able to realize the problems with the traditional cut-and-paste approach for data analysis (i.e., run the analysis separately and copy the results manually), and the advantage of using R Markdown to automatically generate the report.

2.1.3   Personalized mail

One thing you should remember about R Markdown is that you can programmatically generate reports, although most of the time you may be just clicking the `Knit` button in RStudio to generate a single report from a single source document. Being able to program reports is a super power of R Markdown.

Mine Çetinkaya-Rundel once wanted to create personalized handouts for her workshop participants. She used a template R Markdown file, and knitted it in a for-loop to generate 20 PDF files for the 20 participants. Each PDF contained both personalized information and common information. You may read the article `https://rmarkdown.rstudio.com/articles_mail_merge.html` for the technical details.

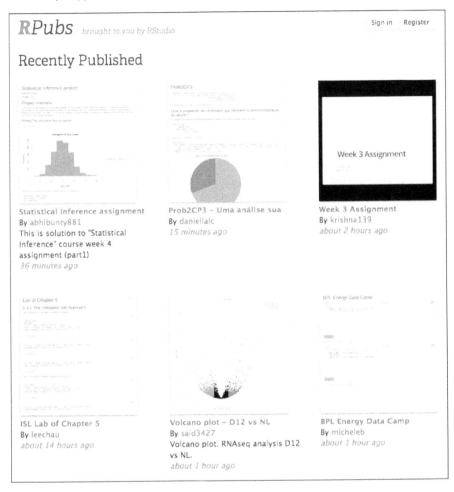

FIGURE 2.3: A screenshot of RPubs.com that contains some homework assginments submitted by students.

2.1.4 2017 Employer Health Benefits Survey

The 2017 Employer Health Benefits Survey[4] was designed and analyzed by the Kaiser Family Foundation, NORC at the University of Chicago, and Health Research & Educational Trust. The full PDF report was written in R Markdown (with the **bookdown** package). It has a unique appearance, which was made possible by heavy customizations in the LaTeX template. This ex-

---

[4]https://www.kff.org/health-costs/report/2017-employer-health-benefits-survey/

ample shows you that if you really care about typesetting, you are free to apply your knowledge about LaTeX to create highly sophisticated reports from R Markdown.

2.1.5   Journal articles

Chris Hartgerink explained how and why he used R Markdown to write dynamic research documents in the post at `https://elifesciences.org/labs/cad57bcf/composing-reproducible-manuscripts-using-r-markdown`. He published a paper titled "Too Good to be False: Nonsignificant Results Revisited" with two co-authors (Hartgerink et al., 2017). The manuscript was written in R Markdown, and results were dynamically generated from the code in R Markdown.

When checking the accuracy of P-values in the psychology literature, his colleagues and he found that P-values could be mistyped or miscalculated, which could lead to inaccurate or even wrong conclusions. If the P-values were dynamically generated and inserted instead of being manually copied from statistical programs, the chance for those problems to exist would be much lower.

Lowndes et al. (2017) also shows that using R Markdown (and version control) not only enhances reproducibility, but also produces better scientific research in less time.

2.1.6   Dashboards at eelloo

R Markdown is used at eelloo (`https://eelloo.nl`) to design and generate research reports. Here is one of their examples (in Dutch): `https://eelloo.nl/groepsrapportages-met-infographics/`, where you can find gauges, bar charts, pie charts, wordclouds, and other types of graphs dynamically generated and embedded in dashboards.

2.1.7   Books

We will introduce the R Markdown extension **bookdown** in Chapter 12. It is an R package that allows you to write books and long-form reports with multiple Rmd files. After this package was published, a large number of books have emerged. You can find a subset of them at `https://bookdown.org`. Some of these books have been printed, and some only have free online versions.

There have also been students who wrote their dissertations/theses with **bookdown**, such as Ed Berry: `https://eddjberry.netlify.com/post/writing-your-thesis-with-bookdown/`. Chester Ismay has even provided an R package **thesisdown** (`https://github.com/ismayc/thesisdown`) that can render a thesis in various formats. Several other people have customized this package for their own institutions, such as Zhian N. Kamvar's **beaverdown** (`https://github.com/zkamvar/beaverdown`) and Ben Marwick's **huskydown** (`https://github.com/benmarwick/huskydown`).

2.1.8 Websites

The **blogdown** package to be introduced in Chapter 10 can be used to build general-purpose websites (including blogs and personal websites) based on R Markdown. You may find tons of examples at `https://github.com/rbind` or by searching on Twitter: `https://twitter.com/search?q=blogdown`. Here are a few impressive websites that I can quickly think of off the top of my head:

- Rob J Hyndman's personal website: `https://robjhyndman.com` (a very comprehensive academic website).

- Amber Thomas's personal website: `https://amber.rbind.io` (a rich project portfolio).

- Emi Tanaka's personal website: `https://emitanaka.github.io` (in particular, check out the beautiful showcase page).

- "Live Free or Dichotomize" by Nick Strayer and Lucy D'Agostino McGowan: `http://livefreeordichotomize.com` (the layout is elegant, and the posts are useful and practical).

2.2 Compile an R Markdown document

The usual way to compile an R Markdown document is to click the `Knit` button as shown in Figure 2.1, and the corresponding keyboard shortcut is `Ctrl + Shift + K` (`Cmd + Shift + K` on macOS). Under the hood, RStudio calls the function `rmarkdown::render()` to render the document *in a new R session*. Please note the emphasis here, which often confuses R Markdown users. Rendering an Rmd document in a new R session means that *none of the objects in your current R session (e.g., those you created in your R console) are available to that session.*[5] Reproducibility is the main reason that RStudio uses a new R session to render your Rmd documents: in most cases, you may want your documents to continue to work the next time you open R, or in other people's computing environments. See this StackOverflow answer[6] if you want to know more.

If you must render a document in the current R session, you can also call `rmarkdown::render()` by yourself, and pass the path of the Rmd file to this function. The second argument of this function is the output format, which defaults to the first output format you specify in the YAML metadata (if it is missing, the default is `html_document`). When you have multiple output formats in the metadata, and do not want to use the first one, you can specify the one you want in the second argument, e.g., for an Rmd document `foo.Rmd` with the metadata:

```
output:
  html_document:
    toc: true
  pdf_document:
    keep_tex: true
```

You can render it to PDF via:

```
rmarkdown::render('foo.Rmd', 'pdf_document')
```

The function call gives you much more freedom (e.g., you can generate a

---

[5]This is not strictly true, but mostly true. You may save objects in your current R session to a file, e.g., `.RData`, and load it in a new R session.

[6]https://stackoverflow.com/a/48494678/559676

series of reports in a loop), but you should bear reproducibility in mind when you render documents this way. Of course, you can start a new and clean R session by yourself, and call `rmarkdown::render()` in that session. As long as you do not manually interact with that session (e.g., manually creating variables in the R console), your reports should be reproducible.

Another main way to work with Rmd documents is the R Markdown Notebooks, which will be introduced in Section 3.2. With notebooks, you can run code chunks individually and see results right inside the RStudio editor. This is a convenient way to interact or experiment with code in an Rmd document, because you do not have to compile the whole document. Without using the notebooks, you can still partially execute code chunks, but the execution only occurs in the R console, and the notebook interface presents results of code chunks right beneath the chunks in the editor, which can be a great advantage. Again, for the sake of reproducibility, you will need to compile the whole document eventually in a clean environment.

Lastly, I want to mention an "unofficial" way to compile Rmd documents: the function `xaringan::inf_mr()`, or equivalently, the RStudio addin "Infinite Moon Reader". Obviously, this requires you to install the **xaringan** package (Xie, 2018g), which is available on CRAN. The main advantage of this way is LiveReload: a technology that enables you to live preview the output as soon as you save the source document, and you do not need to hit the Knit button. The other advantage is that it compiles the Rmd document *in the current R session*, which may or may not be what you desire. Note that this method only works for Rmd documents that output to HTML, including HTML documents and presentations.

A few R Markdown extension packages, such as **bookdown** and **blogdown**, have their own way of compiling documents, and we will introduce them later.

Note that it is also possible to render a series of reports instead of single one from a single R Markdown source document. You can parameterize an R Markdown document, and generate different reports using different parameters. See Chapter 15 for details.

2.3   Cheat sheets

RStudio has created a large number of cheat sheets, including the one-page
R Markdown cheetahs, which are freely available at `https://www.rstudio.`
`com/resources/cheatsheets/`. There is also a more detailed R Markdown
reference guide. Both documents can be used as quick references after you
become more familiar with R Markdown.

2.4   Output formats

There are two types of output formats in the **rmarkdown** package: documents, and presentations. All available formats are listed below:

- `beamer_presentation`
- `github_document`
- `html_document`
- `ioslides_presentation`
- `latex_document`
- `md_document`
- `odt_document`
- `pdf_document`
- `powerpoint_presentation`
- `rtf_document`
- `slidy_presentation`
- `word_document`

We will document these output formats in detail in Chapters 3 and 4. There
are more output formats provided in other extension packages (starting from
Chapter 5). For the output format names in the YAML metadata of an Rmd
file, you need to include the package name if a format is from an extension
package, e.g.,

```
output: tufte::tufte_html
```

If the format is from the **rmarkdown** package, you do not need the rmark-down:: prefix (although it will not hurt).

When there are multiple output formats in a document, there will be a drop-down menu behind the RStudio Knit button that lists the output format names (Figure 2.4).

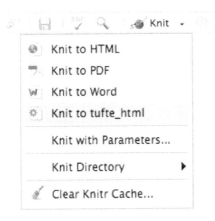

FIGURE 2.4: The output formats listed in the dropdown menu on the RStudio toolbar.

Each output format is often accompanied with several format options. All these options are documented on the R package help pages. For example, you can type ?rmarkdown::html_document in R to open the help page of the html_document format. When you want to use certain options, you have to translate the values from R to YAML, e.g.,

```
html_document(toc = TRUE, toc_depth = 2, dev = 'svg')
```

can be written in YAML as:

```
output:
  html_document:
    toc: true
    toc_depth: 2
    dev: 'svg'
```

The translation is often straightforward. Remember that R's TRUE, FALSE, and NULL are true, false, and null, respectively, in YAML. Character strings in

YAML often do not require the quotes (e.g., dev: 'svg' and dev: svg are the same), unless they contain special characters, such as the colon :. If you are not sure if a string should be quoted or not, test it with the **yaml** package, e.g.,

```
cat(yaml::as.yaml(list(
  title = 'A Wonderful Day',
  subtitle = 'hygge: a quality of coziness'
)))
```

```
title: A Wonderful Day
subtitle: 'hygge: a quality of coziness'
```

Note that the subtitle in the above example is quoted because of the colon.

If a certain option has sub-options (which means the value of this option is a list in R), the sub-options need to be further indented, e.g.,

```
output:
  html_document:
    toc: true
    includes:
      in_header: header.html
      before_body: before.html
```

Some options are passed to **knitr**, such as dev, fig_width, and fig_height. Detailed documentation of these options can be found on the **knitr** documentation page: https://yihui.name/knitr/options/. Note that the actual **knitr** option names can be different. In particular, **knitr** uses . in names, but **rmarkdown** uses _, e.g., fig_width in **rmarkdown** corresponds to fig.width in **knitr**. We apologize for the inconsistencies—programmers often strive for consistencies in their own world, yet one standard plus one standard often equals three standards.[7] If I were to design the **knitr** package again, I would definitely use _.

Some options are passed to Pandoc, such as toc, toc_depth, and number_sections. You should consult the Pandoc documentation when in

---

[7]https://xkcd.com/927/

doubt. R Markdown output format functions often have a `pandoc_args` argument, which should be a character vector of extra arguments to be passed to Pandoc. If you find any Pandoc features that are not represented by the output format arguments, you may use this ultimate argument, e.g.,

```
output:
  pdf_document:
    toc: true
    pandoc_args: ["--wrap=none", "--top-level-division=chapter"]
```

2.5 Markdown syntax

The text in an R Markdown document is written with the Markdown syntax. Precisely speaking, it is Pandoc's Markdown. There are many flavors of Markdown invented by different people, and Pandoc's flavor is the most comprehensive one to our knowledge. You can find the full documentation of Pandoc's Markdown at `https://pandoc.org/MANUAL.html`. We strongly recommend that you read this page at least once to know all the possibilities with Pandoc's Markdown, even if you will not use all of them. This section is adapted from Section 2.1[8] of Xie (2016), and only covers a small subset of Pandoc's Markdown syntax.

2.5.1 Inline formatting

Inline text will be *italic* if surrounded by underscores or asterisks, e.g., `_text_` or `*text*`. **Bold** text is produced using a pair of double asterisks (`**text**`). A pair of tildes (~) turn text to a subscript (e.g., `H~3~PO~4~` renders H_3PO_4). A pair of carets (^) produce a superscript (e.g., `Cu^2+^` renders Cu^{2+}).

To mark text as `inline code`, use a pair of backticks, e.g., `` `code` ``. To include n literal backticks, use at least $n + 1$ backticks outside, e.g., you can use four backticks to preserve three backtick inside: ```` ```` ```code``` ```` ````, which is rendered as ``` ```code``` ```.

---

[8] `https://bookdown.org/yihui/bookdown/markdown-syntax.html`

Hyperlinks are created using the syntax `[text](link)`, e.g., `[RStu-dio](https://www.rstudio.com)`. The syntax for images is similar: just add an exclamation mark, e.g., `![alt text or image title](path/to/image)`. Footnotes are put inside the square brackets after a caret `^[]`, e.g., `^[This is a footnote.]`.

There are multiple ways to insert citations, and we recommend that you use BibTeX databases, because they work better when the output format is La-TeX/PDF. Section 2.8[9] of Xie (2016) has explained the details. The key idea is that when you have a BibTeX database (a plain-text file with the conventional filename extension `.bib`) that contains entries like:

```
@Manual{R-base,
  title = {R: A Language and Environment for Statistical
    Computing},
  author = {{R Core Team}},
  organization = {R Foundation for Statistical Computing},
  address = {Vienna, Austria},
  year = {2017},
  url = {https://www.R-project.org/},
}
```

You may add a field named `bibliography` to the YAML metadata, and set its value to the path of the BibTeX file. Then in Markdown, you may use `@R-base` (which generates "R Core Team (2018)") or `[@R-base]` (which generates "(R Core Team, 2018)") to reference the BibTeX entry. Pandoc will automatically generated a list of references in the end of the document.

2.5.2   Block-level elements

Section headers can be written after a number of pound signs, e.g.,

```
# First-level header
```

```
## Second-level header
```

```
### Third-level header
```

---

[9]`https://bookdown.org/yihui/bookdown/citations.html`

If you do not want a certain heading to be numbered, you can add {-} or
{.unnumbered} after the heading, e.g.,

```
# Preface {-}
```

Unordered list items start with *, -, or +, and you can nest one list within
another list by indenting the sub-list, e.g.,

```
- one item
- one item
- one item
    - one more item
    - one more item
    - one more item
```

The output is:

- one item
- one item
- one item
    - one more item
    - one more item
    - one more item

Ordered list items start with numbers (you can also nest lists within lists),
e.g.,

```
1. the first item
2. the second item
3. the third item
    - one unordered item
    - one unordered item
```

The output does not look too much different with the Markdown source:

1. the first item
2. the second item
3. the third item
    - one unordered item
    - one unordered item

Blockquotes are written after >, e.g.,

```
> "I thoroughly disapprove of duels. If a man should challenge me,
  I would take him kindly and forgivingly by the hand and lead him
  to a quiet place and kill him."
>
> --- Mark Twain
```

The actual output (we customized the style for blockquotes in this book):

"I thoroughly disapprove of duels. If a man should challenge me, I would take him kindly and forgivingly by the hand and lead him to a quiet place and kill him."

— Mark Twain

Plain code blocks can be written after three or more backticks, and you can also indent the blocks by four spaces, e.g.,

```
```
This text is displayed verbatim / preformatted
```
```

```
Or indent by four spaces:

    This text is displayed verbatim / preformatted
```

In general, you'd better leave at least one empty line between adjacent but different elements, e.g., a header and a paragraph. This is to avoid ambiguity to the Markdown renderer. For example, does "#" indicate a header below?

```
In R, the character
# indicates a comment.
```

And does "-" mean a bullet point below?

```
The result of 5
- 3 is 2.
```

Different flavors of Markdown may produce different results if there are no blank lines.

2.5.3   Math expressions

Inline LaTeX equations can be written in a pair of dollar signs using the LaTeX syntax, e.g., `$f(k) = {n \choose k} p^{k} (1-p)^{n-k}$` (actual output: $f(k) = \binom{n}{k} p^k (1-p)^{n-k}$); math expressions of the display style can be written in a pair of double dollar signs, e.g., `$$f(k) = {n \choose k} p^{k} (1-p)^{n-k}$$`, and the output looks like this:

$$f(k) = \binom{n}{k} p^k (1-p)^{n-k}$$

You can also use math environments inside $ $ or $$ $$, e.g.,

```
$$\begin{array}{ccc}
x_{11} & x_{12} & x_{13}\\
x_{21} & x_{22} & x_{23}
\end{array}$$
```

$$\begin{array}{ccc}
x_{11} & x_{12} & x_{13}\\
x_{21} & x_{22} & x_{23}
\end{array}$$

```
$$X = \begin{bmatrix}1 & x_{1}\\
1 & x_{2}\\
1 & x_{3}
\end{bmatrix}$$
```

$$X = \begin{bmatrix} 1 & x_1 \\ 1 & x_2 \\ 1 & x_3 \end{bmatrix}$$

```
$$\Theta = \begin{pmatrix}\alpha & \beta\\
\gamma & \delta
\end{pmatrix}$$
```

$$\Theta = \begin{pmatrix} \alpha & \beta \\ \gamma & \delta \end{pmatrix}$$

```
$$\begin{vmatrix}a & b\\
c & d
\end{vmatrix}=ad-bc$$
```

$$\begin{vmatrix} a & b \\ c & d \end{vmatrix} = ad - bc$$

2.6 R code chunks and inline R code

You can insert an R code chunk either using the RStudio toolbar (the `Insert` button) or the keyboard shortcut `Ctrl` + `Alt` + `I` (`Cmd` + `Option` + `I` on macOS).

There are a lot of things you can do in a code chunk: you can produce text output, tables, or graphics. You have fine control over all these output via chunk options, which can be provided inside the curly braces (between ```{r and }). For example, you can choose hide text output via the chunk option `results` = `'hide'`, or set the figure height to 4 inches via `fig.height` = 4. Chunk options are separated by commas, e.g.,

```
```{r, chunk-label, results='hide', fig.height=4}
```

The value of a chunk option can be an arbitrary R expression, which makes chunk options extremely flexible. For example, the chunk option `eval` controls whether to evaluate (execute) a code chunk, and you may conditionally evaluate a chunk via a variable defined previously, e.g.,
```

```
```{r}
# execute code if the date is later than a specified day
do_it = Sys.Date() > '2018-02-14'
```
```

```
```{r, eval=do_it}
x = rnorm(100)
```
```

There are a large number of chunk options in **knitr** documented at `https://yihui.name/knitr/options`. We list a subset of them below:

- `eval`: Whether to evaluate a code chunk.

- `echo`: Whether to echo the source code in the output document (someone may not prefer reading your smart source code but only results).

- `results`: When set to `'hide'`, text output will be hidden; when set to `'asis'`, text output is written "as-is", e.g., you can write out raw Markdown text from R code (like `cat('**Markdown** is cool.\n')`). By default, text output will be wrapped in verbatim elements (typically plain code blocks).

- `collapse`: Whether to merge text output and source code into a single code block in the output. This is mostly cosmetic: `collapse = TRUE` makes the output more compact, since the R source code and its text output are displayed in a single output block. The default `collapse = FALSE` means R expressions and their text output are separated into different blocks.

- `warning`, `message`, and `error`: Whether to show warnings, messages, and errors in the output document. Note that if you set `error = FALSE`, rmarkdown::render() will halt on error in a code chunk, and the error will be displayed in the R console. Similarly, when `warning = FALSE` or `message = FALSE`, these messages will be shown in the R console.

- `include`: Whether to include anything from a code chunk in the output document. When `include = FALSE`, this whole code chunk is excluded in the output, but note that it will still be evaluated if `eval = TRUE`. When you are trying to set `echo = FALSE`, `results = 'hide'`, `warning = FALSE`, and `message = FALSE`, chances are you simply mean a single option `include = FALSE` instead of suppressing different types of text output individually.

- `cache`: Whether to enable caching. If caching is enabled, the same code chunk will not be evaluated the next time the document is compiled (if the code chunk was not modified), which can save you time. However, I want to honestly remind you of the two hard problems in computer science (via Phil Karlton): naming things, and cache invalidation. Caching can be handy but also tricky sometimes.

- `fig.width` and `fig.height`: The (graphical device) size of R plots in inches. R plots in code chunks are first recorded via a graphical device in **knitr**, and then written out to files. You can also specify the two options together in a single chunk option `fig.dim`, e.g., `fig.dim = c(6, 4)` means `fig.width = 6` and `fig.height = 4`.

- `out.width` and `out.height`: The output size of R plots in the output document. These options may scale images. You can use percentages, e.g., `out.width = '80%'` means 80% of the page width.

- `fig.align`: The alignment of plots. It can be `'left'`, `center`, or `'right'`.

- `dev`: The graphical device to record R plots. Typically it is `'pdf'` for LaTeX output, and `'png'` for HTML output, but you can certainly use other devices, such as `'svg'` or `'jpeg'`.

- `fig.cap`: The figure caption.

- `child`: You can include a child document in the main document. This option takes a path to an external file.

Chunk options in **knitr** can be surprisingly powerful. For example, you can create animations from a series of plots in a code chunk. I will not explain how here because it requires an external software package[10], but encourage you to read the documentation carefully to discover the possibilities. You may also read Xie (2015), which is a comprehensive guide to the **knitr** package, but unfortunately biased towards LaTeX users for historical reasons (which was one of the reasons why I wanted to write this R Markdown book).

There is an optional chunk option that does not take any value, which is the chunk label. It should be the first option in the chunk header. Chunk labels are mainly used in filenames of plots and cache. If the label of a chunk is missing, a default one of the form `unnamed-chunk-i` will be generated, where `i` is incremental. I strongly recommend that you only use alphanumeric characters (`a-z`, `A-Z` and `0-9`) and dashes (`-`) in labels, because they are not spe-

---

[10] https://blogdown-demo.rbind.io/2018/01/31/gif-animations/

cial characters and will surely work for all output formats. Other characters, spaces and underscores in particular, may cause trouble in certain packages, such as **bookdown**.

If a certain option needs to be frequently set to a value in multiple code chunks, you can consider setting it globally in the first code chunk of your document, e.g.,

```
```{r, setup, include=FALSE}
knitr::opts_chunk$set(fig.width = 8, collapse = TRUE)
```
```

Besides code chunks, you can also insert values of R objects inline in text. For example:

```
```{r}
x = 5   # radius of a circle
```
```

```
For a circle with the radius `r x`,
its area is `r pi * x^2`.
```

2.6.1 Figures

By default, figures produced by R code will be placed immediately after the code chunk they were generated from. For example:

```
```{r}
plot(cars, pch = 18)
```
```

You can provide a figure caption using `fig.cap` in the chunk options. If the document output format supports the option `fig_caption: true` (e.g., the output format `rmarkdown::html_document`), the R plots will be placed into figure environments. In the case of PDF output, such figures will be automatically numbered. If you also want to number figures in other formats (such as HTML), please see the **bookdown** package in Chapter 12 (in particular, see Section 12.4.4).

PDF documents are generated through the LaTeX files generated from R Markdown. A highly surprising fact to LaTeX beginners is that figures float by default: even if you generate a plot in a code chunk on the first page, the whole figure environment may float to the next page. This is just how LaTeX works by default. It has a tendency to float figures to the top or bottom of pages. Although it can be annoying and distracting, we recommend that you refrain from playing the "Whac-A-Mole" game in the beginning of your writing, i.e., desparately trying to position figures "correctly" while they seem to be always dodging you. You may wish to fine-tune the positions once the content is complete using the `fig.pos` chunk option (e.g., `fig.pos = 'h'`). See https://www.sharelatex.com/learn/Positioning_images_and_tables for possible values of `fig.pos` and more general tips about this behavior in LaTeX. In short, this can be a difficult problem for PDF output.

To place multiple figures side-by-side from the same code chunk, you can use the `fig.hold='hold'` option along with the `out.width` option. Figure 2.5 shows an example with two plots, each with a width of 50%.

```
par(mar = c(4, 4, 0.2, 0.1))
plot(cars, pch = 19)
plot(pressure, pch = 17)
```

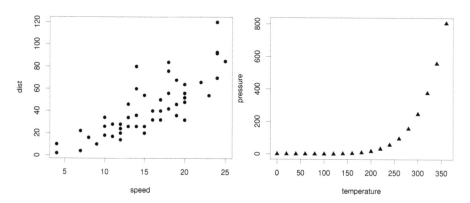

FIGURE 2.5: Two plots side-by-side.

If you want to include a graphic that is not generated from R code, you may use the `knitr::include_graphics()` function, which gives you more control over the attributes of the image than the Markdown syntax of `![alt` text or image title](path/to/image) (e.g., you can specify the image width via out.width). Figure 2.6 provides an example of this.

```{r, out.width='25%', fig.align='center', fig.cap='...'}
knitr::include_graphics('images/hex-rmarkdown.png')
```

FIGURE 2.6: The R Markdown hex logo.

2.6.2 Tables

The easiest way to include tables is by using knitr::kable(), which can create tables for HTML, PDF and Word outputs.[11] Table captions can be included by passing caption to the function, e.g.,

```{r tables-mtcars}
knitr::kable(iris[1:5, ], caption = 'A caption')
```

Tables in non-LaTeX output formats will always be placed after the code block. For LaTeX/PDF output formats, tables have the same issue as figures: they may float. If you want to avoid this behavior, you will need to use the LaTeX package longtable[12], which can break tables across multiple pages. This can be achieved by adding \usepackage{longtable} to your LaTeX preamble, and passing longtable = TRUE to kable().

If you are looking for more advanced control of the styling of tables, you are

---

[11]You may also consider the **pander** package. There are several other packages for producing tables, including **xtable**, **Hmisc**, and **stargazer**, but these are generally less compatible with multiple output formats.

[12]https://www.ctan.org/pkg/longtable

recommended to use the **kableExtra**[13] package, which provides functions to customize the appearance of PDF and HTML tables. Formatting tables can be a very complicated task, especially when certain cells span more than one column or row. It is even more complicated when you have to consider different output formats. For example, it is difficult to make a complex table work for both PDF and HTML output. We know it is disappointing, but sometimes you may have to consider alternative ways of presenting data, such as using graphics.

We explain in Section 12.3 how the **bookdown** package extends the functionality of **rmarkdown** to allow for figures and tables to be easily cross-referenced within your text.

2.7   Other language engines

A less well-known fact about R Markdown is that many other languages are also supported, such as Python, Julia, C++, and SQL. The support comes from the **knitr** package, which has provided a large number of *language engines*. Language engines are essentially functions registered in the object `knitr::knit_engine`. You can list the names of all available engines via:

```
names(knitr::knit_engines$get())
```

```
##  [1] "awk"      "bash"      "coffee"
##  [4] "gawk"     "groovy"    "haskell"
##  [7] "lein"     "mysql"     "node"
## [10] "octave"   "perl"      "psql"
## [13] "Rscript"  "ruby"      "sas"
## [16] "scala"    "sed"       "sh"
## [19] "stata"    "zsh"       "highlight"
## [22] "Rcpp"     "tikz"      "dot"
## [25] "c"        "fortran"   "fortran95"
## [28] "asy"      "cat"       "asis"
## [31] "stan"     "block"     "block2"
## [34] "js"       "css"       "sql"
```

---

[13]https://cran.r-project.org/package=kableExtra

```
## [37] "go"          "python"      "julia"
## [40] "theorem"     "lemma"       "corollary"
## [43] "proposition" "conjecture"  "definition"
## [46] "example"     "exercise"    "proof"
## [49] "remark"      "solution"
```

Most engines have been documented in Chapter 11 of Xie (2015). The engines from theorem to solution are only available when you use the **bookdown** package, and the rest are shipped with the **knitr** package. To use a different language engine, you can change the language name in the chunk header from r to the engine name, e.g.,

````
```{python}
x = 'hello, python world!'
print(x.split(' '))
```
````

For engines that rely on external interpreters such as python, perl, and ruby, the default interpreters are obtained from Sys.which(), i.e., using the interpreter found via the environment variable PATH of the system. If you want to use an alternative interpreter, you may specify its path in the chunk option engine.path. For example, you may want to use Python 3 instead of the default Python 2, and we assume Python 3 is at /usr/bin/python3 (may not be true for your system):

````
```{python, engine.path = '/usr/bin/python3'}
import sys
print(sys.version)
```
````

You can also change the engine interpreters globally for multiple engines, e.g.,

```
knitr::opts_chunk$set(engine.path = list(
  python = '~/anaconda/bin/python',
  ruby = '/usr/local/bin/ruby'
))
```

Note that you can use a named list to specify the paths for different engines.

Most engines will execute each code chunk in a separate new session (via a `system()` call in R), which means objects created in memory in a previous code chunk will not be directly available to latter code chunks. For example, if you create a variable in a `bash` code chunk, you will not be able to use it in the next `bash` code chunk. Currently the only exceptions are `r`, `python`, and `julia`. Only these engines execute code in the same session throughout the document. To clarify, all `r` code chunks are executed in the same R session, all `python` code chunks are executed in the same Python session, and so on, but *the R session and the Python session are independent.*[14]

I will introduce some specific features and examples for a subset of language engines in **knitr** below. Note that most chunk options should work for both R and other languages, such as `eval` and `echo`, so these options will not be mentioned again.

2.7.1 Python

The `python` engine is based on the **reticulate** package (Allaire et al., 2018b), which makes it possible to execute all Python code chunks in the same Python session. If you actually want to execute a certain code chunk in a new Python session, you may use the chunk option `python.reticulate` = `FALSE`. If you are using a **knitr** version lower than 1.18, you should update your R packages.

Below is a relatively simple example that shows how you can create/modify variables, and draw graphics in Python code chunks. Values can be passed to or retrieved from the Python session. To pass a value to Python, assign to `py$name`, where `name` is the variable name you want to use in the Python session; to retrieve a value from Python, also use `py$name`.

```
---
title: "Python code chunks in R Markdown"
date: 2018-02-22
---

## A normal R code chunk
```

---

[14]This is not strictly true, since the Python session is actually launched from R. What I mean here is that you should not expect to use R variables and Python variables interchangeably without explicitly importing/exporting variables between the two sessions.

````
```{r}
library(reticulate)
x = 42
print(x)
```
````

Modify an R variable

In the following chunk, the value of `x` on the right hand side is `r x`, which was defined in the previous chunk.

````
```{r}
x = x + 12
print(x)
```
````

A Python chunk

This works fine and as expected.

````
```{python}
x = 42 * 2
print(x)
```
````

The value of `x` in the Python session is `r py$x`.
It is not the same `x` as the one in R.

Modify a Python variable

````
```{python}
x = x + 18
print(x)
```
````

Retrieve the value of `x` from the Python session again:

````
```{r}
````
````

```
py$x
```

Assign to a variable in the Python session from R:

````
```{r}
py$y = 1:5
```
````

See the value of `y` in the Python session:

````
```{python}
print(y)
```
````

Python graphics

You can draw plots using the **matplotlib** package in Python.

````
```{python}
import matplotlib.pyplot as plt
plt.plot([0, 2, 1, 4])
plt.show()
```
````

You may learn more about the **reticulate** package from https://rstudio.github.io/reticulate/.

2.7.2   Shell scripts

You can also write Shell scripts in R Markdown, if your system can run them (the executable bash or sh should exist). Usually this is not a problem for Linux or macOS users. It is not impossible for Windows users to run Shell scripts, but you will have to install additional software (such as Cygwin[15] or the Linux Subsystem).

---

[15]https://www.cygwin.com

```{bash}
echo "Hello Bash!"
cat flights1.csv flights2.csv flights3.csv > flights.csv
```

Shell scripts are executed via the system2() function in R. Basically **knitr** passes a code chunk to the command bash -c to run it.

2.7.3 SQL

The sql engine uses the **DBI**[16] package to execute SQL queries, print their results, and optionally assign the results to a data frame.

To use the sql engine, you first need to establish a DBI connection to a database (typically via the DBI::dbConnect() function). You can make use of this connection in a sql chunk via the connection option. For example:

```{r}
library(DBI)
db = dbConnect(RSQLite::SQLite(), dbname = "sql.sqlite")
```

```{sql, connection=db}
SELECT * FROM trials
```

By default, SELECT queries will display the first 10 records of their results within the document. The number of records displayed is controlled by the max.print option, which is in turn derived from the global **knitr** option sql.max.print (e.g., knitr::opts_knit$set(sql.max.print = 10); N.B. it is opts_knit instead of opts_chunk). For example, the following code chunk displays the first 20 records:

```{sql, connection=db, max.print = 20}
SELECT * FROM trials
```

---

[16]https://cran.rstudio.com/package=DBI

You can specify no limit on the records to be displayed via max.print = -1
or max.print = NA.

By default, the sql engine includes a caption that indicates the total number
of records displayed. You can override this caption using the tab.cap chunk
option. For example:

```
```{sql, connection=db, tab.cap = "My Caption"}
SELECT * FROM trials
```
```

You can specify that you want no caption all via tab.cap = NA.

If you want to assign the results of the SQL query to an R object as a data
frame, you can do this using the output.var option, e.g.,

```
```{sql, connection=db, output.var="trials"}
SELECT * FROM trials
```
```

When the results of a SQL query are assigned to a data frame, no records will
be printed within the document (if desired, you can manually print the data
frame in a subsequent R chunk).

If you need to bind the values of R variables into SQL queries, you can do so
by prefacing R variable references with a ?. For example:

```
```{r}
subjects = 10
```
```

```
```{sql, connection=db, output.var="trials"}
SELECT * FROM trials WHERE subjects >= ?subjects
```
```

If you have many SQL chunks, it may be helpful to set a default for the connection chunk option in the setup chunk, so that it is not necessary to specify
the connection on each individual chunk. You can do this as follows:

````
```{r setup}
library(DBI)
db = dbConnect(RSQLite::SQLite(), dbname = "sql.sqlite")
knitr::opts_chunk$set(connection = "db")
```
````

Note that the connection option should be a string naming the connection object (not the object itself). Once set, you can execute SQL chunks without specifying an explicit connection:

````
```{sql}
SELECT * FROM trials
```
````

2.7.4 Rcpp

The Rcpp engine enables compilation of C++ into R functions via the **Rcpp** sourceCpp() function. For example:

````
```{Rcpp}
#include <Rcpp.h>
using namespace Rcpp;

// [[Rcpp::export]]
NumericVector timesTwo(NumericVector x) {
  return x * 2;
}
```
````

Executing this chunk will compile the code and make the C++ function timesTwo() available to R.

You can cache the compilation of C++ code chunks using standard **knitr** caching, i.e., add the cache = TRUE option to the chunk:

````
```{Rcpp, cache=TRUE}
#include <Rcpp.h>
````
````

```
using namespace Rcpp;

// [[Rcpp::export]]
NumericVector timesTwo(NumericVector x) {
  return x * 2;
}
```

In some cases, it is desirable to combine all of the Rcpp code chunks in a document into a single compilation unit. This is especially useful when you want to intersperse narrative between pieces of C++ code (e.g., for a tutorial or user guide). It also reduces total compilation time for the document (since there is only a single invocation of the C++ compiler rather than multiple).

To combine all Rcpp chunks into a single compilation unit, you use the ref.label chunk option along with the knitr::all_rcpp_labels() function to collect all of the Rcpp chunks in the document. Here is a simple example:

All C++ code chunks will be combined to the chunk below:

````
```{Rcpp, ref.label=knitr::all_rcpp_labels(), include=FALSE}
```
````

First we include the header `Rcpp.h`:

````
```{Rcpp, eval=FALSE}
#include <Rcpp.h>
```
````

Then we define a function:

````
```{Rcpp, eval=FALSE}
// [[Rcpp::export]]
int timesTwo(int x) {
  return x * 2;
}
```
````

The two Rcpp chunks that include code will be collected and compiled together in the first Rcpp chunk via the ref.label chunk option. Note that we
set the eval = FALSE option on the Rcpp chunks with code in them to prevent
them from being compiled again.

2.7.5 Stan

The stan engine enables embedding of the Stan probabilistic programming
language[17] within R Markdown documents.

The Stan model within the code chunk is compiled into a stanmodel object,
and is assigned to a variable with the name given by the output.var option.
For example:

```
```{stan, output.var="ex1"}
parameters {
  real y[2];
}
model {
  y[1] ~ normal(0, 1);
  y[2] ~ double_exponential(0, 2);
}
```

```{r}
library(rstan)
fit = sampling(ex1)
print(fit)
```
```

2.7.6 JavaScript and CSS

If you are using an R Markdown format that targets HTML output
(e.g., html_document and ioslides_presenation, etc.), you can include
JavaScript to be executed within the HTML page using the JavaScript engine
named js.

---

[17]http://mc-stan.org

For example, the following chunk uses jQuery (which is included in most R Markdown HTML formats) to change the color of the document title to red:

```
```{js, echo=FALSE}
$('.title').css('color', 'red')
```
```

Similarly, you can embed CSS rules in the output document. For example, the following code chunk turns text within the document body red:

```
```{css, echo=FALSE}
body {
  color: red;
}
```
```

Without the chunk option echo = FALSE, the JavaScript/CSS code will be displayed verbatim in the output document, which is probably not what you want.

2.7.7 Julia

The Julia[18] language is supported through the **JuliaCall** package (Li, 2018). Similar to the python engine, the julia engine runs all Julia code chunks in the same Julia session. Below is a minimal example:

```
```{julia}
a = sqrt(2);  # the semicolon inhibits printing
```
```

2.7.8 C and Fortran

For code chunks that use C or Fortran, **knitr** uses R CMD SHLIB to compile the code, and load the shared object (a *.so file on Unix or *.dll on Windows). Then you can use .C() / .Fortran() to call the C / Fortran functions, e.g.,

---

[18] https://julialang.org

```
```{c, test-c, results='hide'}
void square(double *x) {
  *x = *x * *x;
}
```
```

Test the `square()` function:

```
```{r}
.C('square', 9)
.C('square', 123)
```
```

You can find more examples on different language engines in the GitHub repository `https://github.com/yihui/knitr-examples` (look for filenames that contain the word "engine").

2.8 Interactive documents

R Markdown documents can also generate interactive content. There are two types of interactive R Markdown documents: you can use the HTML Widgets framework, or the Shiny framework (or both). They will be described in more detail in Chapter 16 and Chapter 19, respectively.

2.8.1 HTML widgets

The HTML Widgets framework is implemented in the R package **htmlwidgets** (Vaidyanathan et al., 2018), interfacing JavaScript libraries that create interactive applications, such as interactive graphics and tables. Several widget packages have been developed based on this framework, such as **DT** (Xie, 2018c), **leaflet** (Cheng et al., 2018), and **dygraphs** (Vanderkam et al., 2017). Visit `https://www.htmlwidgets.org` to know more about widget packages as well as how to develop a widget package by yourself.

Figure 2.7 shows an interactive map created via the **leaflet** package, and the source document is below:

```
---
title: "An Interactive Map"
---


Below is a map that shows the location of the
Department of Statistics, Iowa State University.

```{r out.width='100%', echo=FALSE}
library(leaflet)
leaflet() %>% addTiles() %>%
  setView(-93.65, 42.0285, zoom = 17) %>%
  addPopups(
    -93.65, 42.0285,
    'Here is the <b>Department of Statistics</b>, ISU'
  )
```
```

Although HTML widgets are based on JavaScript, the syntax to create them in R is often pure R syntax.

If you include an HTML widget in a non-HTML output format, such as a PDF, **knitr** will try to embed a screenshot of the widget if you have installed the R package **webshot** (Chang, 2017) and the PhantomJS package (via `web-shot::install_phantomjs()`).

2.8.2   Shiny documents

The **shiny** package (Chang et al., 2018) builds interactive web apps powered by R. To call Shiny code from an R Markdown document, add `runtime: shiny` to the YAML metadata, like in this document:

```
---
title: "A Shiny Document"
output: html_document
runtime: shiny
---
```

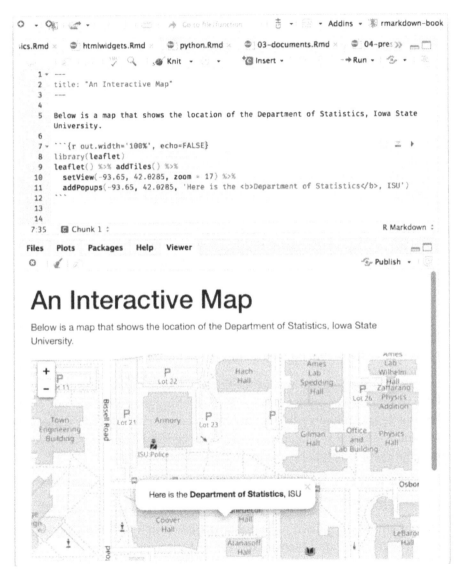

FIGURE 2.7: An R Markdown document with a leaflet map widget.

A standard R plot can be made interactive by wrapping it in the Shiny `renderPlot()` function. The `selectInput()` function creates the input widget to drive the plot.

```
```{r eruptions, echo=FALSE}
selectInput(
  'breaks', label = 'Number of bins:',
  choices = c(10, 20, 35, 50), selected = 20
)

renderPlot({
  par(mar = c(4, 4, .1, .5))
  hist(
    faithful$eruptions, as.numeric(input$breaks),
    col = 'gray', border = 'white',
    xlab = 'Duration (minutes)', main = ''
  )
})
```
```

Figure 2.8 shows the output, where you can see a dropdown menu that allows you to choose the number of bins in the histogram.

You may use Shiny to run any R code that you like in response to user actions. Since web browsers cannot execute R code, Shiny interactions occur on the server side and rely on a live R session. By comparison, HTML widgets do not require a live R session to support them, because the interactivity comes from the client side (via JavaScript in the web browser).

You can learn more about Shiny at `https://shiny.rstudio.com`.

HTML widgets and Shiny elements rely on HTML and JavaScript. They will work in any R Markdown format that is viewed in a web browser, such as HTML documents, dashboards, and HTML5 presentations.

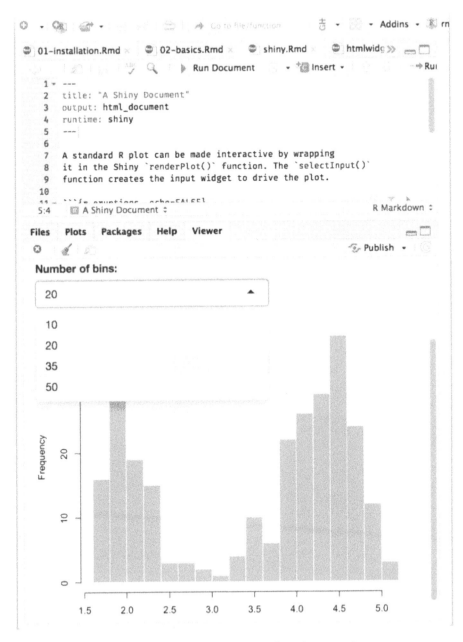

FIGURE 2.8: An R Markdown document with a Shiny widget.

Part II

Output Formats

3

Documents

The very original version of Markdown was invented mainly to write HTML content more easily. For example, you can write a bullet with `-` text instead of the verbose HTML code `<ul><li>text</li></ul>`, or a quote with `>` text instead of `<blockquote>text</blockquote>`.

The syntax of Markdown has been greatly extended by Pandoc. What is more, Pandoc makes it possible to convert a Markdown document to a large variety of output formats. In this chapter, we will introduce the features of various document output formats. In the next two chapters, we will document the presentation formats and other R Markdown extensions, respectively.

3.1 HTML document

As we just mentioned before, Markdown was originally designed for HTML output, so it may not be surprising that the HTML format has the richest features among all output formats. We recommend that you read this full section before you learn other output formats, because other formats have several features in common with the HTML document format, and we will not repeat these features in the corresponding sections.

To create an HTML document from R Markdown, you specify the `html_document` output format in the YAML metadata of your document:

```
---
title: Habits
author: John Doe
date: March 22, 2005
output: html_document
---
```

3.1.1   Table of contents

You can add a table of contents (TOC) using the `toc` option and specify the depth of headers that it applies to using the `toc_depth` option. For example:

```
---
title: "Habits"
output:
  html_document:
    toc: true
    toc_depth: 2
---
```

If the table of contents depth is not explicitly specified, it defaults to 3 (meaning that all level 1, 2, and 3 headers will be included in the table of contents).

3.1.1.1   Floating TOC

You can specify the `toc_float` option to float the table of contents to the left of the main document content. The floating table of contents will always be visible even when the document is scrolled. For example:

```
---
title: "Habits"
output:
  html_document:
    toc: true
    toc_float: true
---
```

You may optionally specify a list of options for the `toc_float` parameter which control its behavior. These options include:

- `collapsed` (defaults to `TRUE`) controls whether the TOC appears with only the top-level (e.g., H2) headers. If collapsed initially, the TOC is automatically expanded inline when necessary.

- `smooth_scroll` (defaults to `TRUE`) controls whether page scrolls are animated when TOC items are navigated to via mouse clicks.

For example:

```
---
title: "Habits"
output:
  html_document:
    toc: true
    toc_float:
      collapsed: false
      smooth_scroll: false
---
```

3.1.2 Section numbering

You can add section numbering to headers using the `number_sections` option:

```
---
title: "Habits"
output:
  html_document:
    toc: true
    number_sections: true
---
```

Note that if you do choose to use the `number_sections` option, you will likely also want to use # (H1) headers in your document as ## (H2) headers will include a decimal point, because without H1 headers, you H2 headers will be numbered with `0.1`, `0.2`, and so on.

3.1.3 Tabbed sections

You can organize content using tabs by applying the `.tabset` class attribute to headers within a document. This will cause all sub-headers of the header with the `.tabset` attribute to appear within tabs rather than as standalone sections. For example:

```
## Quarterly Results {.tabset}
```

```
### By Product

(tab content)

### By Region

(tab content)
```

You can also specify two additional attributes to control the appearance and behavior of the tabs. The `.tabset-fade` attribute causes the tabs to fade in and out when switching between tabs. The `.tabset-pills` attribute causes the visual appearance of the tabs to be "pill" (see Figure 3.1) rather than traditional tabs. For example:

```
## Quarterly Results {.tabset .tabset-fade .tabset-pills}
```

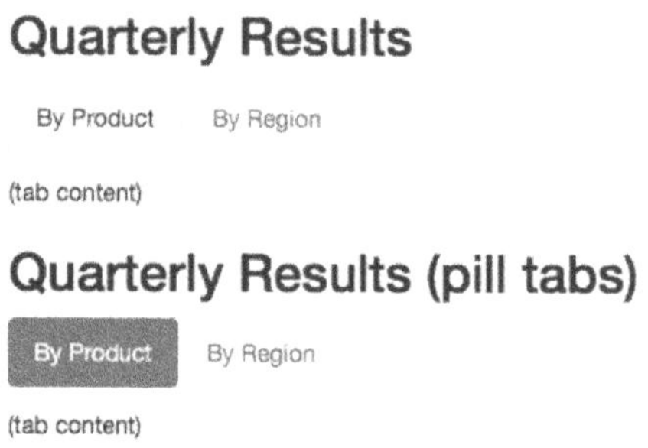

FIGURE 3.1: Traditional tabs and pill tabs on an HTML page.

3.1.4   Appearance and style

There are several options that control the appearance of HTML documents:

- theme specifies the Bootstrap theme to use for the page (themes are drawn from the Bootswatch[1] theme library). Valid themes include default, cerulean, journal, flatly, readable, spacelab, united, cosmo, lumen, paper, sandstone, simplex, and yeti. Pass null for no theme (in this case you can use the css parameter to add your own styles).

---

[1] https://bootswatch.com/3/

- `highlight` specifies the syntax highlighting style. Supported styles include `default`, `tango`, `pygments`, `kate`, `monochrome`, `espresso`, `zenburn`, `haddock`, and `textmate`. Pass `null` to prevent syntax highlighting.

- `smart` indicates whether to produce typographically correct output, converting straight quotes to curly quotes, `---` to em-dashes, `--` to en-dashes, and `...` to ellipses. Note that `smart` is enabled by default.

For example:

```
---
title: "Habits"
output:
  html_document:
    theme: united
    highlight: tango
---
```

3.1.4.1 Custom CSS

You can add your own CSS to an HTML document using the `css` option:

```
---
title: "Habits"
output:
  html_document:
    css: styles.css
---
```

If you want to provide all of the styles for the document from your own CSS you set the `theme` (and potentially `highlight`) to `null`:

```
---
title: "Habits"
output:
  html_document:
    theme: null
    highlight: null
    css: styles.css
---
```

You can also target specific sections of documents with custom CSS by adding ids or classes to section headers within your document. For example the following section header:

```
## Next Steps {#nextsteps .emphasized}
```

Would enable you to apply CSS to all of its content using either of the following CSS selectors:

```
#nextsteps {
    color: blue;
}

.emphasized {
    font-size: 1.2em;
}
```

3.1.5 Figure options

There are a number of options that affect the output of figures within HTML documents:

- `fig_width` and `fig_height` can be used to control the default figure width and height (7x5 is used by default).

- `fig_retina` specifies the scaling to perform for retina displays (defaults to 2, which currently works for all widely used retina displays). Set to `null` to prevent retina scaling.

- `fig_caption` controls whether figures are rendered with captions.

- `dev` controls the graphics device used to render figures (defaults to `png`).

For example:

```
---
title: "Habits"
output:
  html_document:
    fig_width: 7
```

TABLE 3.1: The possible values of the `df_print` option for the `html_document` format.

Option	Description
default	Call the `print.data.frame` generic method
kable	Use the `knitr::kable` function
tibble	Use the `tibble::print.tbl_df` function
paged	Use `rmarkdown::print.paged_df` to create a pageable table

```
    fig_height: 6
    fig_caption: true
---
```

3.1.6 Data frame printing

You can enhance the default display of data frames via the `df_print` option. Valid values are shown in Table 3.1.

3.1.6.1 Paged printing

When the `df_print` option is set to `paged`, tables are printed as HTML tables with support for pagination over rows and columns. For instance (see Figure 3.2):

```
---
title: "Motor Trend Car Road Tests"
output:
  html_document:
    df_print: paged
---

```{r}
mtcars
```
```

Table 3.2 shows the available options for paged tables.

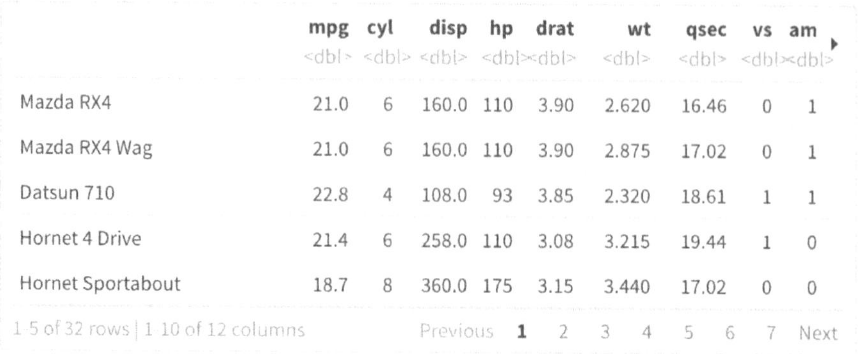

	mpg	cyl	disp	hp	drat	wt	qsec	vs	am	
	<dbl>	<dbl>	<dbl>	<dbl>	<dbl>	<dbl>	<dbl>	<dbl>	<dbl>	
Mazda RX4	21.0	6	160.0	110	3.90	2.620	16.46	0	1	
Mazda RX4 Wag	21.0	6	160.0	110	3.90	2.875	17.02	0	1	
Datsun 710	22.8	4	108.0	93	3.85	2.320	18.61	1	1	
Hornet 4 Drive	21.4	6	258.0	110	3.08	3.215	19.44	1	0	
Hornet Sportabout	18.7	8	360.0	175	3.15	3.440	17.02	0	0	

1 5 of 32 rows | 1 10 of 12 columns   Previous **1**   2   3   4   5   6   7   Next

FIGURE 3.2: A paged table in the HTML output document.

TABLE 3.2: The options for paged HTML tables.

Option	Description
max.print	The number of rows to print.
rows.print	The number of rows to display.
cols.print	The number of columns to display.
cols.min.print	The minimum number of columns to display.
pages.print	The number of pages to display under page navigation.
paged.print	When set to FALSE turns off paged tables.
rownames.print	When set to FALSE turns off row names.

These options are specified in each chunk like below:

```
```{r cols.print=3, rows.print=3}
mtcars
```
```

3.1.7 Code folding

When the **knitr** chunk option echo = TRUE is specified (the default behavior), the R source code within chunks is included within the rendered document. In some cases, it may be appropriate to exclude code entirely (echo = FALSE) but in other cases you might want the code to be available but not visible by default.

The `code_folding: hide` option enables you to include R code but have it hidden by default. Users can then choose to show hidden R code chunks either individually or document wide. For example:

```
---
title: "Habits"
output:
  html_document:
    code_folding: hide
---
```

You can specify `code_folding: show` to still show all R code by default but then allow users to hide the code if they wish.

3.1.8 MathJax equations

By default, MathJax[2] scripts are included in HTML documents for rendering LaTeX and MathML equations. You can use the `mathjax` option to control how MathJax is included:

- Specify `"default"` to use an HTTPS URL from a CDN host (currently provided by RStudio).

- Specify `"local"` to use a local version of MathJax (which is copied into the output directory). Note that when using `"local"` you also need to set the `self_contained` option to `false`.

- Specify an alternate URL to load MathJax from another location.

- Specify `null` to exclude MathJax entirely.

For example, to use a local copy of MathJax:

```
---
title: "Habits"
output:
  html_document:
    mathjax: local
    self_contained: false
---
```

---

[2]https://www.mathjax.org

To use a self-hosted copy of MathJax:

```
---
title: "Habits"
output:
  html_document:
    mathjax: "http://example.com/MathJax.js"
---
```

To exclude MathJax entirely:

```
---
title: "Habits"
output:
  html_document:
    mathjax: null
---
```

3.1.9   Document dependencies

By default, R Markdown produces standalone HTML files with no external dependencies, using data: URIs to incorporate the contents of linked scripts, stylesheets, images, and videos. This means you can share or publish the file just like you share Office documents or PDFs. If you would rather keep dependencies in external files, you can specify self_contained: false. For example:

```
---
title: "Habits"
output:
  html_document:
    self_contained: false
---
```

Note that even for self-contained documents, MathJax is still loaded externally (this is necessary because of its big size). If you want to serve MathJax locally, you should specify mathjax: local and self_contained: false.

One common reason to keep dependencies external is for serving R Markdown documents from a website (external dependencies can be cached separately by browsers, leading to faster page load times). In the case of serving multiple R Markdown documents you may also want to consolidate dependent library files (e.g. Bootstrap, and MathJax, etc.) into a single directory shared by multiple documents. You can use the `lib_dir` option to do this. For example:

```
---
title: "Habits"
output:
  html_document:
    self_contained: false
    lib_dir: libs
---
```

3.1.10 Advanced customization

3.1.10.1 Keeping Markdown

When **knitr** processes an R Markdown input file, it creates a Markdown (`*.md`) file that is subsequently transformed into HTML by Pandoc. If you want to keep a copy of the Markdown file after rendering, you can do so using the `keep_md` option:

```
---
title: "Habits"
output:
  html_document:
    keep_md: true
---
```

3.1.10.2 Includes

You can do more advanced customization of output by including additional HTML content or by replacing the core Pandoc template entirely. To include content in the document header or before/after the document body, you use the `includes` option as follows:

```
---
title: "Habits"
output:
  html_document:
    includes:
      in_header: header.html
      before_body: doc_prefix.html
      after_body: doc_suffix.html
---
```

3.1.10.3 Custom templates

You can also replace the underlying Pandoc template using the `template` option:

```
---
title: "Habits"
output:
  html_document:
    template: quarterly_report.html
---
```

Consult the documentation on Pandoc templates[3] for additional details on templates. You can also study the default HTML template `default.html5`[4] as an example.

3.1.10.4 Markdown extensions

By default, R Markdown is defined as all Pandoc Markdown extensions with the following tweaks for backward compatibility with the old **markdown** package (Allaire et al., 2017):

```
+autolink_bare_uris
+ascii_identifier
+tex_math_single_backslash
```

You can enable or disable Markdown extensions using the `md_extensions`

---

[3]http://pandoc.org/MANUAL.html#templates
[4]https://github.com/jgm/pandoc-templates/

option (you preface an option with - to disable and + to enable it). For example:

```
---
title: "Habits"
output:
  html_document:
    md_extensions: -autolink_bare_uris+hard_line_breaks
---
```

The above would disable the `autolink_bare_uris` extension, and enable the `hard_line_breaks` extension.

For more on available markdown extensions see the Pandoc Markdown specification[5].

3.1.10.5 Pandoc arguments

If there are Pandoc features that you want to use but lack equivalents in the YAML options described above, you can still use them by passing custom `pandoc_args`. For example:

```
---
title: "Habits"
output:
  html_document:
    pandoc_args: [
      "--title-prefix", "Foo",
      "--id-prefix", "Bar"
    ]
---
```

Documentation on all available pandoc arguments can be found in the Pandoc User Guide[6].

---

[5]http://pandoc.org/MANUAL.html#pandocs-markdown
[6]http://pandoc.org/MANUAL.html#options

3.1.11   Shared options

If you want to specify a set of default options to be shared by multiple documents within a directory, you can include a file named `_output.yml` within
the directory. Note that no YAML delimiters (`---`) or the enclosing `output`
field are used in this file. For example:

```
html_document:
  self_contained: false
  theme: united
  highlight: textmate
```

It should not be written as:

```
---
output:
  html_document:
    self_contained: false
    theme: united
    highlight: textmate
---
```

All documents located in the same directory as `_output.yml` will inherit its
options. Options defined explicitly within documents will override those
specified in the shared options file.

3.1.12   HTML fragments

If want to create an HTML fragment rather than a full HTML document you
can use the `html_fragment` format. For example:

```
---
output: html_fragment
---
```

Note that HTML fragments are not complete HTML documents. They do
not contain the standard header content that HTML documents do (they only
contain content in the `<body>` tags of normal HTML documents). They are intended for inclusion within other web pages or content management systems

(like blogs). As such, they do not support features like themes or code high-lighting (it is expected that the environment they are ultimately published within handles these things).

3.2 Notebook

An R Notebook is an R Markdown document with chunks that can be executed independently and interactively, with output visible immediately beneath the input. See Figure 3.3 for an example.

R Notebooks are an implementation of Literate Programming[7] that allows for direct interaction with R while producing a reproducible document with publication-quality output.

Any R Markdown document can be used as a notebook, and all R Notebooks can be rendered to other R Markdown document types. A notebook can therefore be thought of as a special execution mode for R Markdown documents. The immediacy of notebook mode makes it a good choice while authoring the R Markdown document and iterating on code. When you are ready to publish the document, you can share the notebook directly, or render it to a publication format with the Knit button.

3.2.1 Using Notebooks

3.2.1.1 Creating a Notebook

You can create a new notebook in RStudio with the menu command File -> New File -> R Notebook, or by using the html_notebook output type in your document's YAML metadata.

```
---
title: "My Notebook"
output: html_notebook
---
```

---

[7]https://en.wikipedia.org/wiki/Literate_programming

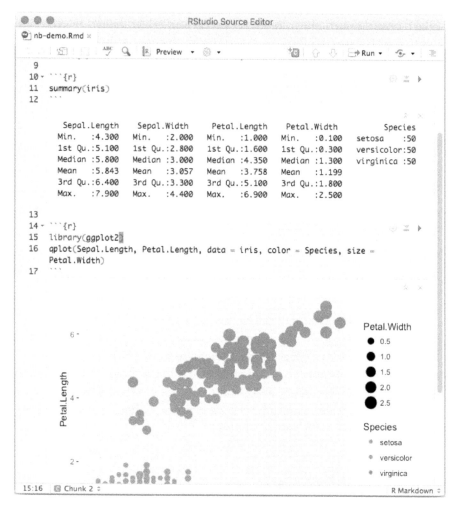

FIGURE 3.3: An R Notebook example.

By default, RStudio enables inline output (Notebook mode) on all R Markdown documents, so you can interact with any R Markdown document as though it were a notebook. If you have a document with which you prefer to use the traditional console method of interaction, you can disable notebook mode by clicking the gear button in the editor toolbar, and choosing Chunk Output in Console (Figure 3.4).

If you prefer to use the console by default for *all* your R Markdown documents (restoring the behavior in previous versions of RStudio), you can make

FIGURE 3.4: Send the R code chunk output to the console.

`Chunk Output in Console` the default: `Tools -> Options -> R Markdown -> Show output inline for all R Markdown documents`.

3.2.1.2   Inserting chunks

Notebook chunks can be inserted quickly using the keyboard shortcut `Ctrl + Alt + I` (macOS: `Cmd + Option + I`), or via the `Insert` menu in the editor toolbar.

Because all of a chunk's output appears beneath the chunk (not alongside the statement which emitted the output, as it does in the rendered R Markdown output), it is often helpful to split chunks that produce multiple outputs into two or more chunks which each produce only one output. To do this, select the code to split into a new chunk (Figure 3.5), and use the same keyboard shortcut for inserting a new code chunk (Figure 3.6).

```
10 ▾ ```{r}
11    plot(data1)
12    plot(data2)
13    ```
```

FIGURE 3.5: Select the code to split into a new chunk.

```
10 ▾ ```{r}
11    plot(data1)
12    ```
13
14 ▾ ```{r}
15    plot(data2)
16    ```
```

FIGURE 3.6: Insert a new chunk from the code selected before.

3.2.1.3 Executing code

Code in the notebook is executed with the same gestures you would use to execute code in an R Markdown document:

1. Use the green triangle button on the toolbar of a code chunk that has the tooltip "Run Current Chunk", or Ctrl + Shift + Enter (macOS: Cmd + Shift + Enter) to run the current chunk.

2. Press Ctrl + Enter (macOS: Cmd + Enter) to run just the current statement. Running a single statement is much like running an entire chunk consisting only of that statement.

3. There are other ways to run a batch of chunks if you click the menu Run on the editor toolbar, such as Run All, Run All Chunks Above, and Run All Chunks Below.

The primary difference is that when executing chunks in an R Markdown document, all the code is sent to the console at once, but in a notebook, only one line at a time is sent. This allows execution to stop if a line raises an error.

There is also a Restart R and Run All Chunks item in the Run menu on the editor toolbar, which gives you a fresh R session prior to running all the chunks. This is similar to the Knit button, which launches a separate R session to compile the document.

When you execute code in a notebook, an indicator will appear in the gutter to show you execution progress (Figure 3.7). Lines of code that have been sent to R are marked with dark green; lines that have not yet been sent to R are marked with light green. If at least one chunk is waiting to be executed, you will see a progress meter appear in the editor's status bar, indicating the number of chunks remaining to be executed. You can click on this meter at any time to jump to the currently executing chunk. When a chunk is waiting to execute, the Run button in its toolbar will change to a "queued" icon. If you do not want the chunk to run, you can click on the icon to remove it from the execution queue.

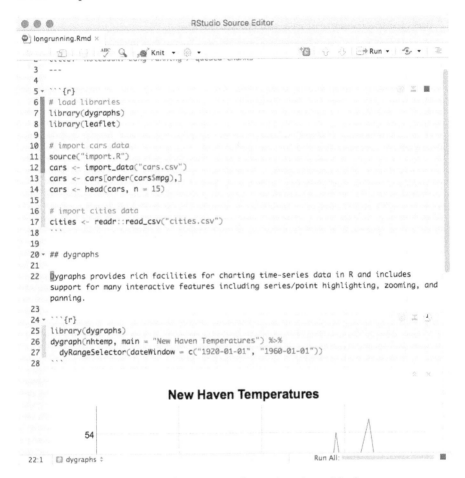

FIGURE 3.7: Insert a new chunk from the code selected before.

In general, when you execute code in a notebook chunk, it will do exactly the same thing as it would if that same code were typed into the console. There are however a few differences:

- **Output**: The most obvious difference is that most forms of output produced from a notebook chunk are shown in the chunk output rather than, for example, the RStudio Viewer or the Plots pane. Console output (including warnings and messages) appears both at the console *and* in the chunk output.

- **Working directory**: The current working directory inside a notebook chunk is always the directory containing the notebook .Rmd file. This makes it easier to use relative paths inside notebook chunks, and also matches the behavior when knitting, making it easier to write code that works identically both interactively and in a standalone render.

  You'll get a warning if you try to change the working directory inside a notebook chunk, and the directory will revert back to the notebook's directory once the chunk is finished executing. You can suppress this warning by using the warnings = FALSE chunk option.

  If it is necessary to execute notebook chunks in a different directory, you can change the working directory for **all** your chunks by using the **knitr** root.dir option. For instance, to execute all notebook chunks in the grandparent folder of the notebook:

  ```r
  knitr::opts_knit$set(root.dir = normalizePath(".."))
  ```

  This option is only effective when used inside the setup chunk. Also note that, as in **knitr**, the root.dir chunk option applies only to chunks; relative paths in Markdown are still relative to the notebook's parent folder.

- **Warnings**: Inside a notebook chunk, warnings are always displayed immediately rather than being held until the end, as in options(warn = 1).

- **Plots**: Plots emitted from a chunk are rendered to match the width of the editor at the time the chunk was executed. The height of the plot is determined by the golden ratio[8]. The plot's display list is saved, too, and the plot is re-rendered to match the editor's width when the editor is resized.

  You can use the fig.width, fig.height, and fig.asp chunk options to manually specify the size of rendered plots in the notebook; you can also use knitr::opts_chunk$set(fig.width = ..., fig.height = ...) in the setup chunk to to set a default rendered size. Note, however, specifying a chunk size manually suppresses the generation of the display list, so plots with manually specified sizes will be resized using simple image scaling when the notebook editor is resized.

To execute an inline R expression in the notebook, put your cursor inside the chunk and press Ctrl + Enter (macOS: Cmd + Enter). As in the execution of ordinary chunks, the content of the expression will be sent to the R console

---

[8]https://en.wikipedia.org/wiki/Golden_ratio

for evaluation. The results will appear in a small pop-up window next to the code (Figure 3.8).

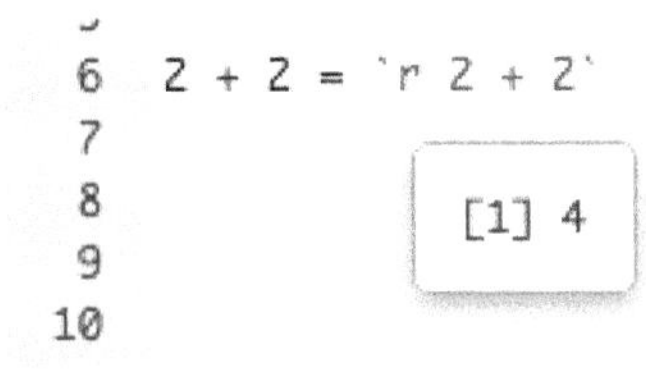

```
 6  2 + 2 = `r 2 + 2`
 7
 8              [1] 4
 9
10
```

FIGURE 3.8: Output from an inline R expression in the notebook.

In notebooks, inline R expressions can only produce text (not figures or other kinds of output). It is also important that inline R expressions executes quickly and do not have side-effects, as they are executed whenever you save the notebook.

Notebooks are typically self-contained. However, in some situations, it is preferable to re-use code from an R script as a notebook chunk, as in **knitr**'s code externalization[9]. This can be done by using `knitr::read_chunk()` in your notebook's setup chunk, along with a special `## ---- chunkname` annotation in the R file from which you intend to read code. Here is a minimal example with two files:

example.Rmd

```
```{r setup}
knitr::read_chunk("example.R")
```
```

example.R

```
## ---- chunk
1 + 1
```

When you execute the empty chunk in the notebook `example.Rmd`, code from the external file `example.R` will be inserted, and the results displayed inline, as though the chunk contained that code (Figure 3.9).

---

[9]`https://yihui.name/knitr/demo/externalization/`

FIGURE 3.9: Execute a code chunk read from an external R script.

3.2.1.4 Chunk output

When code is executed in the notebook, its output appears beneath the code chunk that produced it. You can clear an individual chunk's output by clicking the X button in the upper right corner of the output, or collapse it by clicking the chevron.

It is also possible to clear or collapse all of the output in the document at once using the `Collapse All Output` and `Clear All Output` menu items available on the gear menu in the editor toolbar (Figure 3.4).

If you want to fully reset the state of the notebook, the item `Restart R and Clear Output` on the `Run` menu on the editor toolbar will do the job.

Ordinary R Markdown documents are "knitted", but notebooks are "previewed". While the notebook preview looks similar to a rendered R Markdown document, the notebook preview *does not execute any of your R code chunks*. It simply shows you a rendered copy of the Markdown output of your document along with the most recent chunk output. This preview is generated automatically whenever you save the notebook (whether you are viewing it in RStudio or not); see the section beneath on the `*.nb.html` file for details.

When `html_notebook` is the topmost (default) format in your YAML metadata, you will see a Preview button in the editor toolbar. Clicking it will show you the notebook preview (Figure 3.10).

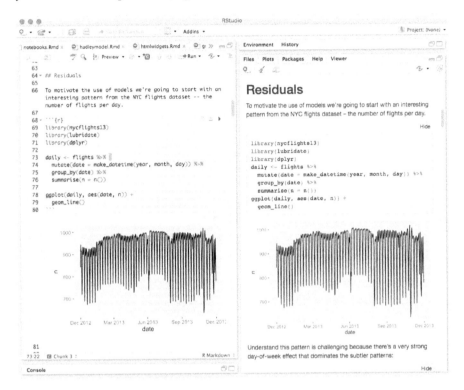

FIGURE 3.10: Preview a notebook.

If you have configured R Markdown previewing to use the Viewer pane (as illustrated in Figure 3.10), the preview will be automatically updated whenever you save your notebook.

When an error occurs while a notebook chunk is executing (Figure 3.11):

1. Execution will stop; the remaining lines of that chunk (and any chunks that have not yet been run) will not be executed.

2. The editor will scroll to the error.

3. The line of code that caused the error will have a red indicator in the editor's gutter.

If you want your notebook to keep running after an error, you can suppress the first two behaviors by specifying error = TRUE in the chunk options.

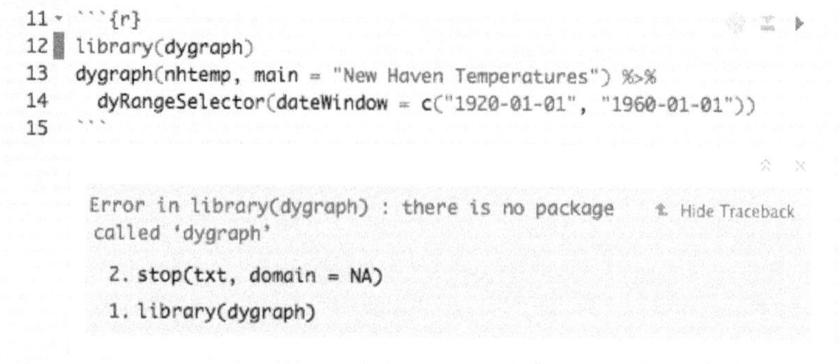

```
11 ▾   ```{r}                                                          ⚒ ✕ ▸
12 ▌  library(dygraph)
13    dygraph(nhtemp, main = "New Haven Temperatures") %>%
14      dyRangeSelector(dateWindow = c("1920-01-01", "1960-01-01"))
15    ```
                                                                        ⚒  ✕

    Error in library(dygraph) : there is no package      ⬆ Hide Traceback
    called 'dygraph'

    2. stop(txt, domain = NA)

    1. library(dygraph)
```

FIGURE 3.11: Errors in a notebook.

In most cases, it should not be necessary to have the console open while using the notebook, as you can see all of the console output in the notebook itself. To preserve vertical space, the console will be automatically collapsed when you open a notebook or run a chunk in the notebook.

If you prefer not to have the console hidden when chunks are executed, uncheck the option from the menu Tools -> Global Options -> R Markdown -> Hide console automatically when executing notebook chunks.

3.2.2   Saving and sharing

3.2.2.1   Notebook file

When a notebook *.Rmd file is saved, a *.nb.html file is created alongside it. This file is a self-contained HTML file which contains both a rendered copy of the notebook with all current chunk outputs (suitable for display on a website) and a copy of the *.Rmd file itself.

You can view the *.nb.html file in any ordinary web browser. It can also be opened in RStudio; when you open there (e.g., using File -> Open File), RStudio will do the following:

1. Extract the bundled *.Rmd file, and place it alongside the *.nb.html file.

2. Open the *.Rmd file in a new RStudio editor tab.

3. Extract the chunk outputs from the *.nb.html file, and place them appropriately in the editor.

Note that the `*.nb.html` file is only created for R Markdown documents that are notebooks (i.e., at least one of their output formats is `html_notebook`). It is possible to have an R Markdown document that includes inline chunk output beneath code chunks, but does not produce an `*.nb.html` file, when `html_notebook` is not specified as an output format for the R Markdown document.

3.2.2.2 Output storage

The document's chunk outputs are also stored in an internal RStudio folder beneath the project's `.Rproj.user` folder. If you work with a notebook but do not have a project open, the outputs are stored in the RStudio state folder in your home directory (the location of this folder varies between the desktop[10] and the server[11]).

3.2.2.3 Version control

One of the major advantages of R Notebooks compared to other notebook systems is that they are plain-text files and therefore work well with version control. We recommend checking in both the `*.Rmd` and `*.nb.html` files into version control, so that both your source code and output are available to collaborators. However, you can choose to include only the `*.Rmd` file (with a `.gitignore` that excludes `*.nb.html`) if you want each collaborator to work with their own private copies of the output.

3.2.3 Notebook format

While RStudio provides a set of integrated tools for authoring R Notebooks, the notebook file format itself is decoupled from RStudio. The **rmarkdown** package provides several functions that can be used to read and write R Notebooks outside of RStudio.

In this section, we describe the internals of the notebook format. It is primarily intended for front-end applications using or embedding R, or other users who are interested in reading and writing documents using the R Notebook format. We recommend that beginners skip this section when reading this book or using notebooks for the first time.

---

[10]https://support.rstudio.com/hc/en-us/articles/200534577
[11]https://support.rstudio.com/hc/en-us/articles/218730228

R Notebooks are HTML documents with data written and encoded in such a way that:

1. The source Rmd document can be recovered, and

2. Chunk outputs can be recovered.

To generate an R Notebook, you can use `rmarkdown::render()` and specify the `html_notebook` output format in your document's YAML metadata. Documents rendered in this form will be generated with the `.nb.html` file extension, to indicate that they are HTML notebooks.

To ensure chunk outputs can be recovered, the elements of the R Markdown document are enclosed with HTML comments, providing more information on the output. For example, chunk output might be serialized in the form:

```
<!-- rnb-chunk-begin -->
<!-- rnb-output-begin -->
<pre><code>Hello, World!</code></pre>
<!-- rnb-output-end -->
<!-- rnb-chunk-end -->
```

Because R Notebooks are just HTML documents, they can be opened and viewed in any web browser; in addition, hosting environments can be configured to recover and open the source Rmd document, and also recover and display chunk outputs as appropriate.

3.2.3.1 Generating R Notebooks with custom output

It is possible to render an HTML notebook with custom chunk outputs inserted in lieu of the result that would be generated by evaluating the associated R code. This can be useful for front-end editors that show the output of chunk execution inline, or for conversion programs from other notebook formats where output is already available from the source format. To facilitate this, one can provide a custom "output source" to `rmarkdown::render()`. Let's investigate with a simple example:

```
rmd_stub = "examples/r-notebook-stub.Rmd"
cat(readLines(rmd_stub), sep = "\n")
```

```
---
title: "R Notebook Stub"
output: html_notebook
---
```

````
```{r chunk-one}
print("Hello, World!")
```
````

Let's try to render this document with a custom output source, so that we can inject custom output for the single chunk within the document. The output source function will accept:

- code: The code within the current chunk.

- context: An environment containing active chunk options and other chunk information.

- ...: Optional arguments reserved for future expansion.

In particular, the context elements label and chunk.index can be used to help identify which chunk is currently being rendered.

```
output_source = function(code, context, ...) {
  logo = file.path(R.home(), "doc/html/logo.jpg")
  if (context$label == "chunk-one") list(
    rmarkdown::html_notebook_output_code("# R Code"),
    paste("Custom output for chunk:", context$chunk.index),
    rmarkdown::html_notebook_output_code("# R Logo"),
    rmarkdown::html_notebook_output_img(logo)
  )
}
```

We can pass our output_source along as part of the output_options list to rmarkdown::render().

```
output_file = rmarkdown::render(
  rmd_stub,
  output_options = list(output_source = output_source),
```

```
    quiet = TRUE
)
```

We have now generated an R Notebook. Open this document[12] in a web browser, and it will show that the output_source function has effectively side-stepped evaluation of code within that chunk, and instead returned the injected result.

3.2.3.2 Implementing output sources

In general, you can provide regular R output in your output source function, but **rmarkdown** also provides a number of endpoints for insertion of custom HTML content. These are documented within ?html_notebook_output.

Using these functions ensures that you produce an R Notebook that can be opened in R frontends (e.g., RStudio).

3.2.3.3 Parsing R Notebooks

The rmarkdown::parse_html_notebook() function provides an interface for recovering and parsing an HTML notebook.

```
parsed = rmarkdown::parse_html_notebook(output_file)
str(parsed, width = 60, strict.width = "wrap")
```

```
List of 4
$ source : chr [1:1759] "<!DOCTYPE html>" "" "<html
    xmlns=\"http://www.w3.org/1999/xhtml\">" "" ...
$ rmd : chr [1:8] "---" "title: \"R Notebook Stub\""
    "output: html_notebook" "---" ...
$ header : chr [1:1598] "<head>" "" "<meta
    charset=\"utf-8\" />" "<meta http-equiv=\"Content-Type\"
    content=\"text/html; charset=utf-8\" />" ...
$ annotations:List of 12
..$ :List of 4
.. ..$ row : int 1701
.. ..$ label: chr "text"
```

---

[12]https://rmarkdown.rstudio.com/notebook/r-notebook-stub.nb.html

```
.. ..$ state: chr "begin"
.. ..$ meta : NULL
..$ :List of 4
.. ..$ row : int 1702
.. ..$ label: chr "text"
.. ..$ state: chr "end"
.. ..$ meta : NULL
..$ :List of 4
.. ..$ row : int 1703
.. ..$ label: chr "chunk"
.. ..$ state: chr "begin"
.. ..$ meta : NULL
..$ :List of 4
.. ..$ row : int 1704
.. ..$ label: chr "source"
.. ..$ state: chr "begin"
.. ..$ meta :List of 1
.. .. ..$ data: chr "```r\n# R Code\n```"
..$ :List of 4
.. ..$ row : int 1706
.. ..$ label: chr "source"
.. ..$ state: chr "end"
.. ..$ meta : NULL
..$ :List of 4
.. ..$ row : int 1707
.. ..$ label: chr "output"
.. ..$ state: chr "begin"
.. ..$ meta :List of 1
.. .. ..$ data: chr "Custom output for chunk: 1\n"
..$ :List of 4
.. ..$ row : int 1709
.. ..$ label: chr "output"
.. ..$ state: chr "end"
.. ..$ meta : NULL
..$ :List of 4
.. ..$ row : int 1710
.. ..$ label: chr "source"
.. ..$ state: chr "begin"
.. ..$ meta :List of 1
.. .. ..$ data: chr "```r\n# R Logo\n```"
```

```
..$ :List of 4
.. ..$ row : int 1712
.. ..$ label: chr "source"
.. ..$ state: chr "end"
.. ..$ meta : NULL
..$ :List of 4
.. ..$ row : int 1713
.. ..$ label: chr "plot"
.. ..$ state: chr "begin"
.. ..$ meta : NULL
..$ :List of 4
.. ..$ row : int 1715
.. ..$ label: chr "plot"
.. ..$ state: chr "end"
.. ..$ meta : NULL
..$ :List of 4
.. ..$ row : int 1716
.. ..$ label: chr "chunk"
.. ..$ state: chr "end"
.. ..$ meta : NULL
```

This interface can be used to recover the original Rmd source, and also (with some more effort from the front-end) the ability to recover chunk outputs from the document itself.

3.3   PDF document

To create a PDF document from R Markdown, you specify the `pdf_document` output format in the YAML metadata:

```
---
title: "Habits"
author: John Doe
date: March 22, 2005
output: pdf_document
---
```

Within R Markdown documents that generate PDF output, you can use raw LaTeX, and even define LaTeX macros. See Pandoc's documentation on the raw_tex extension[13] for details.

Note that PDF output (including Beamer slides) requires an installation of LaTeX (see Chapter 1).

3.3.1 Table of contents

You can add a table of contents using the toc option and specify the depth of headers that it applies to using the toc_depth option. For example:

```
---
title: "Habits"
output:
  pdf_document:
    toc: true
    toc_depth: 2
---
```

If the TOC depth is not explicitly specified, it defaults to 3 (meaning that all level 1, 2, and 3 headers will be included in the TOC).

You can add section numbering to headers using the number_sections option:

```
---
title: "Habits"
output:
  pdf_document:
    toc: true
    number_sections: true
---
```

If you are familiar with LaTeX, number_sections: true means \section{}, and number_sections: false means \section*{} for sections in LaTeX (it also applies to other levels of "sections" such as \chapter{}, and \subsection{}).

---

[13]http://pandoc.org/MANUAL.html#extension-raw_tex

3.3.2   Figure options

There are a number of options that affect the output of figures within PDF documents:

- `fig_width` and `fig_height` can be used to control the default figure width and height (6x4.5 is used by default).

- `fig_crop` controls whether the `pdfcrop` utility, if available in your system, is automatically applied to PDF figures (this is `true` by default). If your graphics device is `postscript`, we recommend that you disable this feature (see more info in this **knitr** issue[14]).

- `fig_caption` controls whether figures are rendered with captions (this is `false` by default).

- `dev` controls the graphics device used to render figures (defaults to `pdf`).

For example:

```
---
title: "Habits"
output:
  pdf_document:
    fig_width: 7
    fig_height: 6
    fig_caption: true
---
```

3.3.3   Data frame printing

You can enhance the default display of data frames via the `df_print` option. Valid values are presented in Table 3.3.

For example:

```
---
title: "Habits"
output:
  pdf_document:
```

---

[14]`https://github.com/yihui/knitr/issues/1365`

TABLE 3.3: The possible values of the `df_print` option for the `pdf_document` format.

Option	Description
default	Call the `print.data.frame` generic method
kable	Use the `knitr::kable()` function
tibble	Use the `tibble::print.tbl_df()` function

```
    df_print: kable
---
```

3.3.4 Syntax highlighting

The `highlight` option specifies the syntax highlighting style. Its usage in `pdf_document` is the same as `html_document` (Section 3.1.4). For example:

```
---
title: "Habits"
output:
  pdf_document:
    highlight: tango
---
```

3.3.5 LaTeX options

Many aspects of the LaTeX template used to create PDF documents can be customized using *top-level* YAML metadata (note that these options do not appear underneath the `output` section, but rather appear at the top level along with `title`, `author`, and so on). For example:

```
---
title: "Crop Analysis Q3 2013"
output: pdf_document
fontsize: 11pt
```

TABLE 3.4: Available top-level YAML metadata variables for LaTeX output.

Variable	Description
lang	Document language code
fontsize	Font size (e.g., `10pt`, `11pt`, or `12pt`)
documentclass	LaTeX document class (e.g., `article`)
classoption	Options for documentclass (e.g., `oneside`)
geometry	Options for geometry class (e.g., `margin=1in`)
mainfont, sansfont, monofont, mathfont	Document fonts (works only with `xelatex` and `lualatex`)
linkcolor, urlcolor, citecolor	Color for internal, external, and citation links

```
geometry: margin=1in
---
```

A few available metadata variables are displayed in Table 3.4 (consult the Pandoc manual for the full list):

3.3.6 LaTeX packages for citations

By default, citations are processed through `pandoc-citeproc`, which works for all output formats. For PDF output, sometimes it is better to use LaTeX packages to process citations, such as `natbib` or `biblatex`. To use one of these packages, just set the option `citation_package` to be `natbib` or `biblatex`, e.g.

```
---
output:
  pdf_document:
    citation_package: natbib
---
```

3.3.7 Advanced customization

3.3.7.1 LaTeX engine

By default, PDF documents are rendered using `pdflatex`. You can specify an alternate engine using the `latex_engine` option. Available engines are `pdflatex`, `xelatex`, and `lualatex`. For example:

```
---
title: "Habits"
output:
  pdf_document:
    latex_engine: xelatex
---
```

The main reasons you may want to use `xelatex` or `lualatex` are: (1) They support Unicode better; (2) It is easier to make use of system fonts. See some posts on Stack Overflow for more detailed explanations, e.g., `https://tex.stackexchange.com/q/3393/9128` and `https://tex.stackexchange.com/q/36/9128`.

3.3.7.2 Keeping intermediate TeX

R Markdown documents are converted to PDF by first converting to a TeX file and then calling the LaTeX engine to convert to PDF. By default, this TeX file is removed, however if you want to keep it (e.g., for an article submission), you can specify the `keep_tex` option. For example:

```
---
title: "Habits"
output:
  pdf_document:
    keep_tex: true
---
```

3.3.7.3 Includes

You can do more advanced customization of PDF output by including additional LaTeX directives and/or content or by replacing the core Pandoc template entirely. To include content in the document header or before/after the document body, you use the `includes` option as follows:

```
---
title: "Habits"
output:
  pdf_document:
    includes:
      in_header: preamble.tex
      before_body: doc-prefix.tex
      after_body: doc-suffix.tex
---
```

3.3.7.4   Custom templates

You can also replace the underlying Pandoc template using the `template` option:

```
---
title: "Habits"
output:
  pdf_document:
    template: quarterly-report.tex
---
```

Consult the documentation on Pandoc templates[15] for additional details on templates. You can also study the default LaTeX template[16] as an example.

3.3.8   Other features

Similar to HTML documents, you can enable or disable certain Markdown extensions for generating PDF documents. See Section 3.1.10.4 for details. You can also pass more custom Pandoc arguments through the `pandoc_args` option (Section 3.1.10.5), and define shared options in `_output.yml` (Section 3.1.11).

---

[15]https://pandoc.org/README.html#templates
[16]https://github.com/jgm/pandoc-templates/blob/master/default.latex

3.4 Word document

To create a Word document from R Markdown, you specify the `word_document` output format in the YAML metadata of your document:

```
---
title: "Habits"
author: John Doe
date: March 22, 2005
output: word_document
---
```

The most notable feature of Word documents is the Word template, which is also known as the "style reference document". You can specify a document to be used as a style reference in producing a `*.docx` file (a Word document). This will allow you to customize things such as margins and other formatting characteristics. For best results, the reference document should be a modified version of a `.docx` file produced using **rmarkdown** or Pandoc. The path of such a document can be passed to the `reference_docx` argument of the `word_document` format. Pass `"default"` to use the default styles. For example:

```
---
title: "Habits"
output:
  word_document:
    reference_docx: my-styles.docx
---
```

For more on how to create and use a reference document, you may watch this short video: `https://vimeo.com/110804387`, or read this detailed article: `https://rmarkdown.rstudio.com/articles_docx.html`.

3.4.1 Other features

Refer to Section 3.1 for the documentation of most features of Word documents, including figure options (Section 3.1.5), data frame printing (Section 3.1.6), syntax highlighting (Section 3.1.4), keeping Markdown (Section 3.1.10.1), Markdown extensions (Section 3.1.10.4), Pandoc arguments (Section 3.1.10.5), and shared options (Section 3.1.11).

3.5 OpenDocument Text document

To create an OpenDocument Text (ODT) document from R Markdown, you specify the `odt_document` output format in the YAML metadata of your document:

```
---
title: "Habits"
author: John Doe
date: March 22, 2005
output: odt_document
---
```

Similar to `word_document`, you can also provide a style reference document to `odt_document` throught the `reference_odt` option. For best results, the reference ODT document should be a modified version of an ODT file produced using **rmarkdown** or Pandoc. For example:

```
---
title: "Habits"
output:
  odt_document:
    reference_odt: my-styles.odt
---
```

3.5.1 Other features

Refer to Section 3.1 for the documentation of most features of ODT documents, including figure options (Section 3.1.5), keeping Markdown (Section 3.1.10.1), header and before/after body inclusions (Section 3.1.10.2), custom templates (Section 3.1.10.3), Markdown extensions (Section 3.1.10.4), Pandoc arguments (Section 3.1.10.5), and shared options (Section 3.1.11).

3.6 Rich Text Format document

To create a Rich Text Format (RTF) document from R Markdown, you specify the rtf_document output format in the YAML metadata of your document:

```
---
title: "Habits"
author: John Doe
date: March 22, 2005
output: rtf_document
---
```

If you know the RTF format really well, you can actually embed raw RTF content in R Markdown. For example, you may create a table in RTF using other software packages, and insert it to your final RTF output document. An RTF document is essentially a plain-text document, so you can read it into R using functions like readLines(). Now suppose you have an RTF table in the file table.rtf. To embed it in R Markdown, you read it and pass to knitr::raw_output(), e.g.,

```
```{r, echo=FALSE}
knitr::raw_output(readLines('table.rtf'))
```
```

3.6.1 Other features

Refer to Section 3.1 for the documentation of most features of RTF documents, including table of contents (Section 3.1.1), figure options (Section 3.1.5), keeping Markdown (Section 3.1.10.1), Markdown extensions (Section 3.1.10.4), Pandoc arguments (Section 3.1.10.5), and shared options (Section 3.1.11).

3.7   Markdown document

In some cases, you might want to produce plain Markdown output from R Markdown (e.g., to create a document for a system that accepts Markdown input like Stack Overflow[17]).

To create a Markdown document from R Markdown, you specify the md_document output format in the front-matter of your document:

```
---
title: "Habits"
author: John Doe
date: March 22, 2005
output: md_document
---
```

3.7.1   Markdown variants

By default, the md_document format produces "strict" Markdown (i.e., conforming to the original Markdown specification with no extensions). You can generate a different flavor of Markdown using the variant option. For example:

```
---
output:
  md_document:
    variant: markdown_github
---
```

Valid values are:

- markdown (Full Pandoc Markdown)
- markdown_strict (Original Markdown specification; the default)
- markdown_github (GitHub Flavored Markdown)
- markdown_mmd (MultiMarkdown)
- markdown_phpextra (PHP Markdown extra)

---

[17]https://stackoverflow.com/editing-help

TABLE 3.5: Markdown variants for some popular publishing systems.

System	Markdown Variant
GitHub Wikis	`markdown_github`
Drupal	`markdown_phpextra`
WordPress.com	`markdown_phpextra+backtick_code_blocks`
StackOverflow	`markdown_strict+autolink_bare_uris`

You can also compose custom Markdown variants. For example:

```
---
output:
  md_document:
    variant: markdown_strict+backtick_code_blocks+autolink_bare_uris
---
```

See Pandoc's Manual for all of the Markdown extensions and their names to be used in composing custom variants.

3.7.1.1 Publishing formats

Many popular publishing systems now accept Markdown as input. Table 3.5 shows the correct Markdown variants to use for some popular systems.

In many cases, you can simply copy and paste the Markdown generated by `rmarkdown::render()` into the editing interface of the system you are targeting. Note, however, that if you have embedded plots or other images, you will need to upload them separately and fix up their URLs to point to the uploaded location. If you intend to build websites based on R Markdown, we recommend that you use the more straightforward solutions such as **blogdown** (Xie et al., 2017; Xie, 2018a) as introduced in Section 10 instead of manually copying the Markdown content.

3.7.2 Other features

Refer to Section 3.1 for the documentation of other features of Markdown documents, including table of contents (Section 3.1.1), figure options (Section

3.1.5), header and before/after body inclusions (Section 3.1.10.2), Pandoc arguments (Section 3.1.10.5), and shared options (Section 3.1.11).

3.8   R package vignette

The `html_vignette` format provides a lightweight alternative to `html_document` suitable for inclusion in packages to be released to CRAN. It reduces the size of a basic vignette from 600Kb to around 10Kb. The format differs from a conventional HTML document as follows:

- Never uses retina figures
- Has a smaller default figure size
- Uses a custom lightweight CSS stylesheet

To use `html_vignette`, you specify it as the output format and add some additional vignette related settings via the `\Vignette*{}` macros:

```
---
title: "Your Vignette Title"
output: rmarkdown::html_vignette
vignette: >
  %\VignetteEngine{knitr::rmarkdown}
  %\VignetteIndexEntry{Your Vignette Title}
  %\VignetteEncoding{UTF-8}
---
```

Note that you should change the `title` field and the `\VignetteIndexEntry{}` to match the title of your vignette.

Most options for `html_document` (Section 3.1) also work for `html_vignette`, except `fig_retina` and `theme`, which have been set to `null` internally in this format.

The `html_vignette` template includes a basic CSS theme. To override this theme, you can specify your own CSS in the document metadata as follows:

```
output:
```

```
rmarkdown::html_vignette:
  css: mystyles.css
```

The default figure size is 3x3. Because the figure width is small, usually you will be able to put two images side-by-side if you set the chunk option `fig.show='hold'`, e.g.,

```
```{r, fig.show='hold'}
plot(1:10)
plot(10:1)
```
```

If you want larger figure sizes you can change the `fig_width` and `fig_height` in the document output options or alternatively override the default options on a per-chunk basis.

4

Presentations

For documents, the basic units are often sections. For presentations, the basic units are slides. A section in the Markdown source document often indicates a new slide in the presentation formats. In this chapter, we introduce the built-in presentation formats in the **rmarkdown** package.

4.1 ioslides presentation

To create an ioslides presentation from R Markdown, you specify the `ioslides_presentation` output format in the YAML metadata of your document. You can create a slide show broken up into sections by using the `#` and `##` heading tags (you can also create a new slide without a header using a horizontal rule (`---`). For example here is a simple slide show (see Figure 4.1 for two sample slides):

```
---
title: "Habits"
author: John Doe
date: March 22, 2005
output: ioslides_presentation
---

# In the morning

## Getting up

- Turn off alarm
- Get out of bed
```

```
## Breakfast

- Eat eggs
- Drink coffee

# In the evening

## Dinner

- Eat spaghetti
- Drink wine

---

```{r, cars, fig.cap="A scatterplot.", echo=FALSE}
plot(cars)
```

Going to sleep

- Get in bed
- Count sheep
```

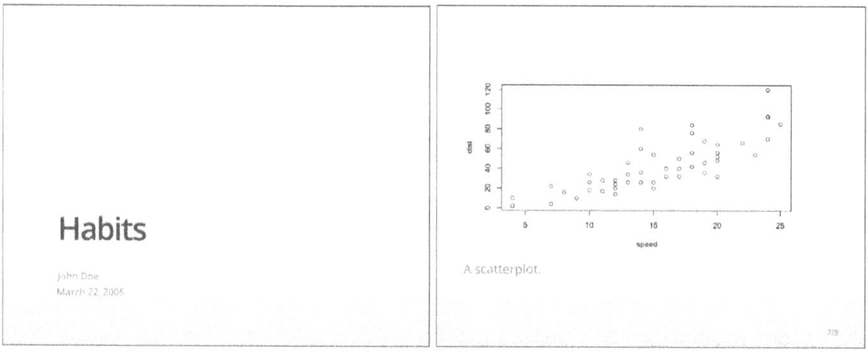

FIGURE 4.1: Two sample slides in an ioslides presentation.

You can add a subtitle to a slide or section by including text after the pipe (|) character. For example:

```
## Getting up | What I like to do first thing
```

4.1.1   Display modes

The following single character keyboard shortcuts enable alternate display modes:

- 'f': enable fullscreen mode
- 'w': toggle widescreen mode
- 'o': enable overview mode
- 'h': enable code highlight mode
- 'p': show presenter notes

Pressing Esc exits all of these modes. See the sections below on *Code Highlighting* and *Presenter Mode* for additional detail on those modes.

4.1.2   Incremental bullets

You can render bullets incrementally by adding the incremental option:

```
---
output:
  ioslides_presentation:
    incremental: true
---
```

If you want to render bullets incrementally for some slides but not others you can (ab)use this syntax for blockquotes:

```
> - Eat eggs
> - Drink coffee
```

4.1.3   Visual appearance

4.1.3.1   Presentation size

You can display the presentation using a wider form factor using the `widescreen` option. You can specify that smaller text be used with the `smaller` option. For example:

```
---
output:
  ioslides_presentation:
    widescreen: true
    smaller: true
---
```

You can also enable the `smaller` option on a slide-by-slide basis by adding the `.smaller` attribute to the slide header:

```
## Getting up {.smaller}
```

4.1.3.2   Transition speed

You can customize the speed of slide transitions using `transition` option. This can be "default", "slower", "faster", or a numeric value with a number of seconds (e.g., `0.5`). For example:

```
---
output:
  ioslides_presentation:
    transition: slower
---
```

4.1.3.3   Build slides

Slides can also have a `.build` attribute that indicate that their content should be displayed incrementally. For example:

```
## Getting up {.build}
```

Slide attributes can be combined if you need to specify more than one. For example:

```
## Getting up {.smaller .build}
```

4.1.3.4   Background images

You can specify a background image for a slide using the attribute `data-background`, and use other attributes including `data-background-size`, `data-background-position`, and `data-background-repeat` to tweak the style of the image. You need to be familiar with CSS to fully understand these four attributes, and they correspond to the CSS properties[1] `background`, `background-size`, `background-position`, and `background-repeat`, respectively. For example:

```
## Getting up {data-background=foo.png data-background-size=cover}
```

4.1.3.5   Custom CSS

You can add your own CSS to an ioslides presentation using the `css` option:

```
---
output:
  ioslides_presentation:
    css: styles.css
---
```

You can also target specific slides or classes of slice with custom CSS by adding IDs or classes to the slides headers within your document. For example the following slide header:

```
## Future Steps {#future-steps .emphasized}
```

---

[1] https://www.w3schools.com/cssref/css3_pr_background.asp

Would enable you to apply CSS to all of its content using either of the following CSS selectors:

```
#future-steps {
    color: blue;
}

.emphasized {
    font-size: 1.2em;
}
```

4.1.4 Code highlighting

It is possible to select subsets of code for additional emphasis by adding a special "highlight" comment around the code. For example:

```
### <b>
x <- 10
y <- x * 2
### </b>
```

The highlighted region will be displayed with a bold font. When you want to help the audience focus exclusively on the highlighted region press the h key and the rest of the code will fade away.

4.1.5 Adding a logo

You can add a logo to the presentation using the `logo` option (by default, the logo will be displayed in a 85x85 pixel square). For example:

```
---
output:
  ioslides_presentation:
    logo: logo.png
---
```

The logo graphic will be rescaled to 85x85 (if necessary) and added to the title slide. A smaller icon version of the logo will be included in the bottom-left footer of each slide.

The logo on the title page and the rectangular element containing it can be customised with CSS. For example:

```
.gdbar img {
  width: 300px !important;
  height: 150px !important;
  margin: 8px 8px;
}

.gdbar {
  width: 400px !important;
  height: 170px !important;
}
```

These selectors are to be placed in the CSS text file.

Similarly, the logo in the footer of each slide can be resized to any desired size. For example:

```
slides > slide:not(.nobackground):before {
  width: 150px;
  height: 75px;
  background-size: 150px 75px;
}
```

This will make the footer logo 150 by 75 pixels in size.

4.1.6 Tables

The ioslides template has an attractive default style for tables so you should not hesitate to add tables for presenting more complex sets of information. Pandoc Markdown supports several syntaxes for defining tables, which are described in the Pandoc Manual.

4.1.7   Advanced layout

You can center content on a slide by adding the `.flexbox` and `.vcenter` attributes to the slide title. For example:

```
## Dinner {.flexbox .vcenter}
```

You can horizontally center content by enclosing it in a `div` tag with class `centered`. For example:

```
<div class="centered">
This text is centered.
</div>
```

You can do a two-column layout using the `columns-2` class. For example:

```
<div class="columns-2">
  ![](image.png)

  - Bullet 1
  - Bullet 2
  - Bullet 3
</div>
```

Note that content will flow across the columns so if you want to have an image on one side and text on the other you should make sure that the image has sufficient height to force the text to the other side of the slide.

4.1.8   Text color

You can color content using base color classes `red`, `blue`, `green`, `yellow`, and `gray` (or variations of them, e.g., `red2`, `red3`, `blue2`, `blue3`, etc.). For example:

```
<div class="red2">
This text is red
</div>
```

4.1.9 Presenter mode

A separate presenter window can also be opened (ideal for when you are presenting on one screen but have another screen that is private to you). The window stays in sync with the main presentation window and also shows presenter notes and a thumbnail of the next slide. To enable presenter mode add `?presentme=true` to the URL of the presentation. For example:

```
my-presentation.html?presentme=true
```

The presenter mode window will open and will always re-open with the presentation until it is disabled with:

```
my-presentation.html?presentme=false
```

To add presenter notes to a slide, you include it within a "notes" `div`. For example:

```
<div class="notes">
This is my *note*.

- It can contain markdown
- like this list
</div>
```

4.1.10 Printing and PDF output

You can print an ioslides presentation from within browsers that have good support for print CSS (as of this writing, Google Chrome has the best support). Printing maintains most of the visual styles of the HTML version of the presentation.

To create a PDF version of a presentation, you can use the menu `Print to PDF` from Google Chrome.

4.1.11 Custom templates

You can replace the underlying Pandoc template using the `template` option:

```
---
title: "Habits"
output:
  ioslides_presentation:
    template: quarterly-report.html
---
```

However, please note that the level of customization that can be achieved is limited compared to the templates of other output formats, because the slides are generated by custom formatting written in Lua, and as such the template used must include the string `RENDERED_SLIDES` as can be found in the default template file with the path `rmarkdown:::rmarkdown_system_file("rmd/ioslides/default.html")`[2].

4.1.12   Other features

Refer to Section 3.1 for the documentation of other features of ioslides presentations, including figure options (Section 3.1.5), MathJax equations (Section 3.1.8), data frame printing (Section 3.1.6), Markdown extensions (Section 3.1.10.4), keeping Markdown (Section 3.1.10.1), document dependencies (Section 3.1.9), header and before/after body inclusions (Section 3.1.10.2), Pandoc arguments (Section 3.1.10.5), and shared options (Section 3.1.11).

4.2   Slidy presentation

To create a Slidy[3] presentation from R Markdown, you specify the `slidy_presentation` output format in the YAML metadata of your document. You can create a slide show broken up into sections by using the `##` heading tag (you can also create a new slide without a header using a horizontal rule (`---`). For example, here is a simple slide show (see Figure 4.2 for two sample slides):

---

[2]`https://github.com/rstudio/rmarkdown/blob/master/inst/rmd/ioslides/default.html`

[3]`https://www.w3.org/Talks/Tools/Slidy2/`

```
---
title: "Habits"
author: John Doe
date: March 22, 2005
output: slidy_presentation
---

# In the morning

## Getting up

- Turn off alarm
- Get out of bed

## Breakfast

- Eat eggs
- Drink coffee

# In the evening

## Dinner

- Eat spaghetti
- Drink wine

---

```{r, cars, fig.cap="A scatterplot.", echo=FALSE}
plot(cars)
```

Going to sleep

- Get in bed
- Count sheep
```

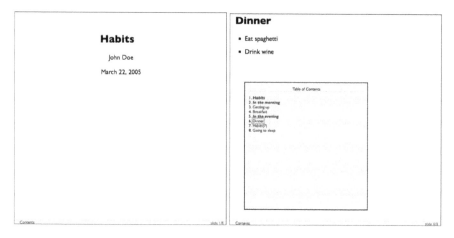

FIGURE 4.2: Two sample slides in a Slidy presentation.

4.2.1 Display modes

The following single character keyboard shortcuts enable alternate display modes:

- 'C': Show table of contents (the right sub-figure in Figure 4.2 has shown the table of contents).
- 'F': Toggles the display of the footer.
- 'A': Toggles display of current vs all slides (useful for printing handouts).
- 'S': Make fonts smaller.
- 'B': Make fonts larger .

4.2.2 Text size

You can use the font_adjustment option to increase or decrease the default font size (e.g., -1 or +1) for the entire presentation. For example:

```
---
output:
  slidy_presentation:
    font_adjustment: -1
---
```

If you want to decrease the text size on an individual slide you can use the .smaller slide attribute. For example:

```
## Getting up {.smaller}
```

If you want to increase the text size on an individual slide you can use the .bigger slide attribute. For example:

```
## Getting up {.bigger}
```

You can also manually adjust the font size during the presentation using the 'S' (smaller) and 'B' (bigger) keys.

4.2.3 Footer elements

You can add a countdown timer to the footer of your slides using the duration option (duration is specified in minutes). For example:

```
---
output:
  slidy_presentation:
    duration: 45
---
```

You can also add custom footer text (e.g., organization name and/or copyright) using the footer option. For example:

```
---
output:
  slidy_presentation:
    footer: "Copyright (c) 2014, RStudio"
---
```

4.2.4 Other features

Refer to Section 3.1 for the documentation of other features of Slidy presentations, including figure options (Section 3.1.5), appearance and style (Section 3.1.4), MathJax equations (Section 3.1.8), data frame printing (Section 3.1.6), Markdown extensions (Section 3.1.10.4), keeping Markdown (Section 3.1.10.1), document dependencies (Section 3.1.9), header and before/after body inclusions (Section 3.1.10.2), custom templates (Section 3.1.10.3), Pandoc arguments (Section 3.1.10.5), and shared options (Section 3.1.11).

Slidy presentations have several features in common with ioslides presentations in Section 4.1. For incremental bullets, see Section 4.1.2. For custom CSS, see Section 4.1.3.5. For printing Slidy slides to PDF, see Section 4.1.10.

4.3   Beamer presentation

To create a Beamer presentation from R Markdown, you specify the `beamer_presentation` output format in the YAML metadata of your document. You can create a slide show broken up into sections by using the # and ## heading tags (you can also create a new slide without a header using a horizontal rule (---). For example, here is a simple slide show (see Figure 4.3 for two sample slides):

```
---
title: "Habits"
author: John Doe
date: March 22, 2005
output: beamer_presentation
---


# In the morning

## Getting up

- Turn off alarm
- Get out of bed

## Breakfast

- Eat eggs
- Drink coffee

# In the evening
```

```
## Dinner

- Eat spaghetti
- Drink wine

---

```{r, cars, fig.cap="A scatterplot.", echo=FALSE}
plot(cars)
```

Going to sleep

- Get in bed
- Count sheep
```

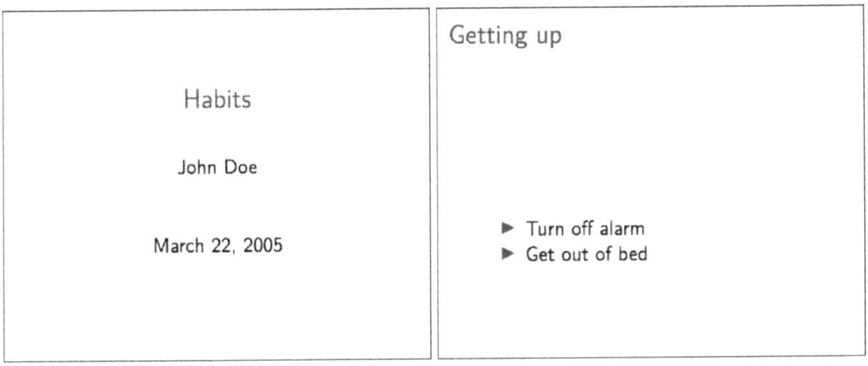

FIGURE 4.3: Two sample slides in a Beamer presentation.

Within R Markdown documents that generate PDF output, you can use raw LaTeX and even define LaTeX macros. See Pandoc's manual for details.

4.3.1 Themes

You can specify Beamer themes using the theme, colortheme, and fonttheme options. For example:

```
---
output:
  beamer_presentation:
    theme: "AnnArbor"
    colortheme: "dolphin"
    fonttheme: "structurebold"
---
```

Figure 4.4 shows two sample slides of the `AnnArbor` theme in the above example. You can find a list of possible themes and color themes at `https:
//hartwork.org/beamer-theme-matrix/`.

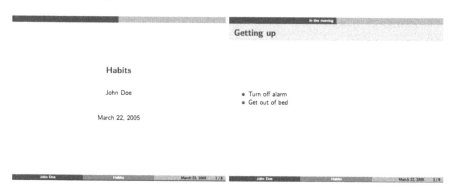

FIGURE 4.4: Two sample slides with the AnnArbor theme in Beamer.

4.3.2   Slide level

The `slide_level` option defines the heading level that defines individual
slides. By default, this is the highest header level in the hierarchy that is
followed immediately by content, and not another header, somewhere in
the document. This default can be overridden by specifying an explicit
`slide_level`:

```
---
output:
  beamer_presentation:
    slide_level: 2
---
```

4.3.3 Other features

Refer to Section 3.1 for the documentation of other features of Beamer presentations, including table of contents (Section 3.1.1), figure options (Section 3.1.5), appearance and style (Section 3.1.4), data frame printing (Section 3.1.6), Markdown extensions (Section 3.1.10.4), header and before/after body inclusions (Section 3.1.10.2), custom templates (Section 3.1.10.3), Pandoc arguments (Section 3.1.10.5), and shared options (Section 3.1.11).

Beamer presentations have a few features in common with ioslides presentations in Section 4.1 and PDF documents in Section 3.3. For incremental bullets, see Section 4.1.2. For how to keep the intermediate LaTeX output file, see Section 3.3.7.2.

4.4 PowerPoint presentation

To create a PowerPoint presentation from R Markdown, you specify the `powerpoint_presentation` output format in the YAML metadata of your document. Please note that this output format is only available in **rmarkdown** >= v1.9, and requires at least Pandoc v2.0.5. You can check the versions of your **rmarkdown** package and Pandoc with `packageVersion('rmarkdown')` and `rmarkdown::pandoc_version()` in R, respectively. The RStudio version 1.1.x ships Pandoc 1.19.2.1, which is not sufficient to generate PowerPoint presentations. You need to either install Pandoc 2.x by yourself if you use RStudio 1.1.x, or install a preview version of RStudio[4] (>= 1.2.633), which has bundled Pandoc 2.x.

Below is a quick example (see Figure 4.5 for a sample slide):

```
---
title: "Habits"
author: John Doe
date: March 22, 2005
output: powerpoint_presentation
---
```

---

[4]https://www.rstudio.com/products/rstudio/download/preview/

```
# In the morning

## Getting up

- Turn off alarm
- Get out of bed

## Breakfast

- Eat eggs
- Drink coffee

# In the evening

## Dinner

- Eat spaghetti
- Drink wine


---


```{r, cars, fig.cap="A scatterplot.", echo=FALSE}
plot(cars)
```


Going to sleep

- Get in bed
- Count sheep
```

The default slide level (i.e., the heading level that defines individual slides) is determined in the same way as in Beamer slides (Section 4.3.2), and you can specify an explicit level via the slide_level option under powerpoint_presentation. You can also start a new slide without a header using a horizontal rule ---.

You can generate most elements supported by Pandoc's Markdown (Section 2.5) in PowerPoint output, such as bold/italic text, footnotes, bullets, LaTeX math expressions, images, and tables, etc.

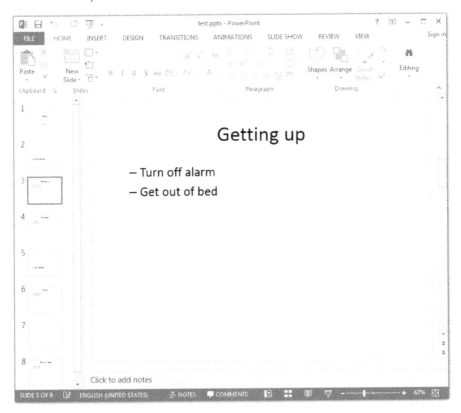

FIGURE 4.5: A sample slide in a PowerPoint presentation.

Please note that images and tables will always be placed on new slides. The only elements that can coexist with an image or table on a slide are the slide header and image/table caption. When you have a text paragraph and an image on the same slide, the image will be moved to a new slide automatically. Images will be scaled automatically to fit the slide, and if the automatic size does not work well, you may manually control the image sizes: for static images included via the Markdown syntax , you may use the width and/or height attributes in a pair of curly braces after the image, e.g., ![caption](foo.png){width=40%}; for images generated dynamically from R code chunks, you can use the chunk options fig.width and fig.height to control the sizes.

Please read the section "Producing slide shows with Pandoc" in Pandoc's manual for more information on slide shows, such as the multi-column layout:

```
:::::: {.columns}
::: {.column width="40%"}
Content of the left column.
:::

::: {.column width="60%"}
Content of the right column.
:::
::::::
```

4.4.1   Custom templates

Like Word documents (Section 3.4), you can customize the appearance of PowerPoint presentations by passing a custom reference document via the reference_doc option, e.g.,

```
---
title: "Habits"
output:
  powerpoint_presentation:
    reference_doc: my-styles.pptx
---
```

Note that the reference_doc option requires a version of **rmarkdown** higher than 1.19:

```
if (packageVersion("rmarkdown") <= "1.9") {
  install.packages("rmarkdown")  # update rmarkdown from CRAN
}
```

Basically any template included in a recent version of Microsoft PowerPoint should work. You can create a new *.pptx file from the PowerPoint menu File -> New with your desired template, save the new file, and use it as the reference document (template) through the reference_doc option. Pandoc will read the styles in the template and apply them to the PowerPoint presentation to be created from R Markdown.

4.4.2 Other features

Refer to Section 3.1 for the documentation of other features of PowerPoint presentations, including table of contents (Section 3.1.1), figure options (Section 3.1.5), data frame printing (Section 3.1.6), keeping Markdown (Section 3.1.10.1), Markdown extensions (Section 3.1.10.4), Pandoc arguments (Section 3.1.10.5), and shared options (Section 3.1.11). As of Pandoc 2.2.1, incremental slides in PowerPoint are not supported yet.

Part III

Extensions

5

Dashboards

R Markdown is customizable and extensible. In Chapters 3 and 4, we have introduced basic document and presentation formats in the **rmarkdown** package, and explained how to customize them. From this chapter on, we will show several more existing extension packages that bring different styles, layouts, and applications to the R Markdown ecosystem. In this chapter, we introduce dashboards based on the **flexdashboard** package (Borges and Allaire, 2017).

Dashboards are particularly common in business-style reports. They can be used to highlight brief and key summaries of a report. The layout of a dashboard is often grid-based, with components arranged in boxes of various sizes.

With the **flexdashboard** package, you can

- Use R Markdown to publish a group of related data visualizations as a dashboard.

- Embed a wide variety of components including HTML widgets, R graphics, tabular data, gauges, value boxes, and text annotations.

- Specify row or column-based layouts (components are intelligently re-sized to fill the browser and adapted for display on mobile devices).

- Create story boards for presenting sequences of visualizations and related commentary.

- Optionally use Shiny to drive visualizations dynamically.

To author a dashboard, you create an R Markdown document with the `flex-dashboard::flex_dashboard` output format. You can also create a document from within RStudio using the `File -> New File -> R Markdown` dialog, and choosing a "Flex Dashboard" template.

If you are not using RStudio, you can create a new `flexdashboard` R Markdown file from the R console:

```
rmarkdown::draft(
  "dashboard.Rmd", template = "flex_dashboard",
  package = "flexdashboard"
)
```

The full documentation of **flexdashboard** is at https://rmarkdown. rstudio.com/flexdashboard/. We will only cover some basic features and usage in this chapter. Dashboards have many features in common with HTML documents (Section 3.1), such as figure options (Section 3.1.5), appearance and style (Section 3.1.4), MathJax equations (Section 3.1.8), header and before/after body inclusions (Section 3.1.10.2), and Pandoc arguments (Section 3.1.10.5), and so on. We also recommend that you take a look at the R help page ?flexdashboard::flex_dashboard to learn about more features and options.

5.1   Layout

The overall rule about the dashboard layout is that a first-level section generates a page, a second-level section generates a column (or a row), and a third-level section generates a box (that contains one or more dashboard components). Below is a quick example:

```
---
title: "Get Started"
output: flexdashboard::flex_dashboard
---

```{r setup, include=FALSE}
library(flexdashboard)
```

Column 1

Chart A
```

```
```{r}
```

Column 2

Chart B

```
```{r}
```

Chart C

```
```{r}
```

Note that a series of dashes under a line of text is the alternative Markdown syntax for the second-level section header, i.e.,

Column 1
----

is equivalent to

Column 1

We used a series of dashes just to make the second-level sections stand out in the source document. By default, the second-level sections generate columns on a dashboard, and the third level sections will be stacked vertically inside columns. You do not have to have columns on a dashboard: when all you have are the third-level sections in the source document, they will be stacked vertically as one column in the output.

The text of the second-level headers will not be displayed in the output. The second-level headers are for the sole purpose of layout, so the actual content of the headers does not matter at all. By contrast, the first-level and third-level headers will be displayed as titles.

Figure 5.1 shows the output of the above example, in which you can see two
```

columns, with the first column containing "Chart A", and the second column containing "Chart B" and "Chart C". We did not really include any R code in the code chunks, so all boxes are empty. In these code chunks, you may write arbitrary R code that generates R plots, HTML widgets, and various other components to be introduced in Section 5.2.

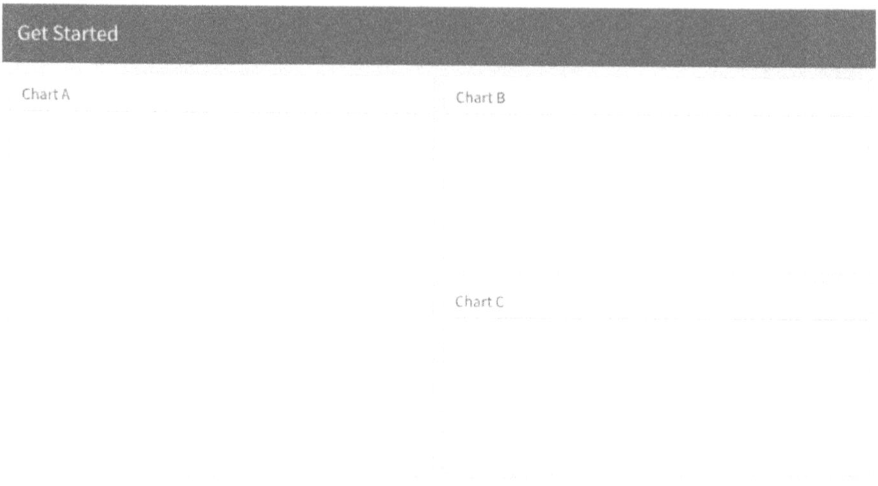

FIGURE 5.1: A quick example of the dashboard layout.

5.1.1   Row-based layouts

You may change the column-oriented layout to the row-oriented layout through the orientation option, e.g.,

```
output:
  flexdashboard::flex_dashboard:
    orientation: rows
```

That means the second-level sections will be rows, and the third-level sections will be arranged as columns within rows.

5.1.2   Attributes on sections

The second-level section headers may have attributes on them, e.g., you can set the width of a column to 350:

```
A narrow column {data-width=350}
--------------------------------
```

For the row-oriented layout, you can set the `data-height` attribute for rows. The `{.tabset}` attribute can be applied on a column so that the third-level sections will be arranged in tabs, e.g.,

```
Two tabs {.tabset}
------------------


### Tab A


### Tab B
```

5.1.3 Multiple pages

When you have multiple first-level sections in the document, they will be displayed as separate pages on the dashboard. Below is an example, and Figure 5.2 shows the output. Note that a series of equal signs is the alternative Markdown syntax for the first-level section headers (you can use a single pound sign #, too).

```
---
title: "Multiple Pages"
output: flexdashboard::flex_dashboard
---


Visualizations {data-icon="fa-signal"}
======================================


### Chart 1

```{r}
```


Chart 2
```

```{r}
```

```
Tables {data-icon="fa-table"}
=====================================
```

```
### Table 1
```

```{r}
```

```
### Table 2
```

```{r}
```

Page titles are displayed as a navigation menu at the top of the dashboard. In this example, we applied icons to page titles through the `data-icon` attribute. You can find other available icons from `https://fontawesome.com`.

FIGURE 5.2: Multiple pages on a dashboard.

5.1.4 Story boards

Besides the column and row-based layouts, you may present a series of visualizations and related commentary through the "storyboard" layout. To enable this layout, you use the option storyboard. Below is an example, and Figure 5.3 shows the output, in which you can see left/right navigation buttons at the top to help you go through all visualizations and associated commentaries one by one.

```
---
title: "Storyboard Commentary"
output:
  flexdashboard::flex_dashboard:
    storyboard: true
---


### A nice scatterplot here

```{r}
plot(cars, pch = 20)
grid()
```

Some commentary about Frame 1.

A beautiful histogram on this board

```{r}
hist(faithful$eruptions, col = 'gray', border = 'white', main = '')
```

Some commentary about Frame 2.
```

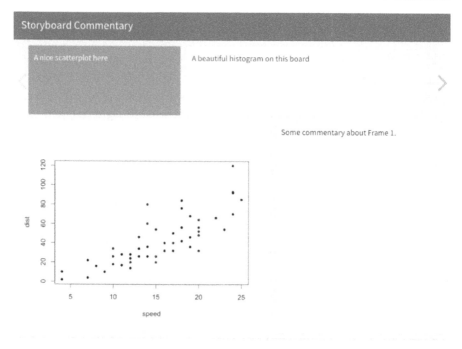

FIGURE 5.3: An example story board.

5.2   Components

A wide variety of components can be included in a dashboard layout, including:

1. Interactive JavaScript data visualizations based on HTML widgets.

2. R graphical output including base, lattice, and grid graphics.

3. Tabular data (with optional sorting, filtering, and paging).

4. Value boxes for highlighting important summary data.

5. Gauges for displaying values on a meter within a specified range.

6. Text annotations of various kinds.

7. A navigation bar to provide more links related to the dashboard.

The first three components work in most R Markdown documents regardless of output formats. Only the latter four are specific to dashboards, and we briefly introduce them in this section.

5.2.1 Value boxes

Sometimes you want to include one or more simple values within a dashboard. You can use the valueBox() function in the **flexdashboard** package to display single values along with a title and an optional icon. For example, here are three side-by-side sections, each displaying a single value (see Figure 5.4 for the output):

```
---
title: "Dashboard Value Boxes"
output:
  flexdashboard::flex_dashboard:
    orientation: rows
---
```

```
```{r setup, include=FALSE}
library(flexdashboard)
# these computing functions are only toy examples
computeArticles = function(...) return(45)
computeComments = function(...) return(126)
computeSpam = function(...) return(15)
```
```

```
### Articles per Day
```

```
```{r}
articles = computeArticles()
valueBox(articles, icon = "fa-pencil")
```
```

```
### Comments per Day
```

```
```{r}
comments = computeComments()
valueBox(comments, icon = "fa-comments")
```
```

```
` ` `
```

Spam per Day

```{r}
spam = computeSpam()
valueBox(
  spam, icon = "fa-trash",
  color = ifelse(spam > 10, "warning", "primary")
)
```
```
` ` `
```

FIGURE 5.4: Three value boxes side by side on a dashboard.

The valueBox() function is called to emit a value and specify an icon.

The third code chunk ("Spam per Day") makes the background color of the value box dynamic using the color parameter. Available colors include "primary", "info", "success", "warning", and "danger" (the default is "primary"). You can also specify any valid CSS color (e.g., "#ffffff", "rgb(100, 100, 100)", etc.).

5.2.2 Gauges

Gauges display values on a meter within a specified range. For example, here is a set of three gauges (see Figure 5.5 for the output):

```
---
title: "Dashboard Gauges"
output:
  flexdashboard::flex_dashboard:
    orientation: rows
---
```

```{r setup, include=FALSE}
library(flexdashboard)
```

Contact Rate

```{r}
gauge(91, min = 0, max = 100, symbol = '%', gaugeSectors(
  success = c(80, 100), warning = c(40, 79), danger = c(0, 39)
))
```

Average Rating

```{r}
gauge(37.4, min = 0, max = 50, gaugeSectors(
  success = c(41, 50), warning = c(21, 40), danger = c(0, 20)
))
```

Cancellations

```{r}
gauge(7, min = 0, max = 10, gaugeSectors(
  success = c(0, 2), warning = c(3, 6), danger = c(7, 10)
))
```

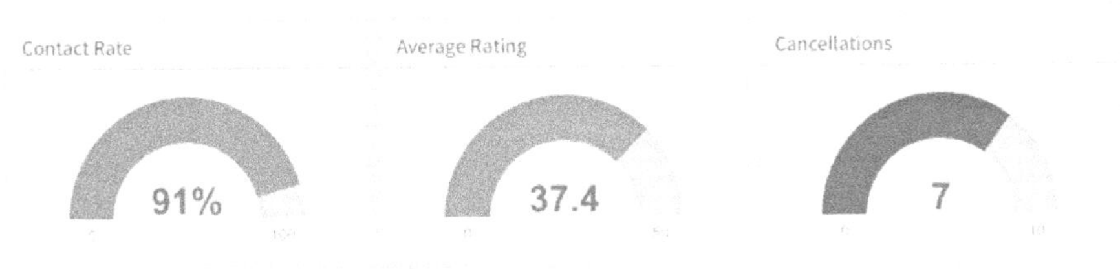

FIGURE 5.5: Three gauges side by side on a dashboard.

There are a few things to note about this example:

1. The gauge() function is used to output a gauge. It has three re-

quired arguments: `value`, `min`, and `max` (these can be any numeric values).

2. You can specify an optional `symbol` to be displayed alongside the value (in the example "%" is used to denote a percentage).

3. You can specify a set of custom color "sectors" using the `gaugeSectors()` function. By default, the current theme's "success" color (typically green) is used for the gauge color. The `sectors` option enables you to specify a set of three value ranges (`success`, `warning`, and `danger`), which cause the gauge's color to change based on its value.

5.2.3 Text annotations

If you need to include additional narrative or explanation within your dashboard, you can do so in the following ways:

1. You can include content at the top of the page before dashboard sections are introduced.

2. You can define dashboard sections that do not include a chart but rather include arbitrary content (text, images, and equations, etc.).

For example, the following dashboard includes some content at the top and a dashboard section that contains only text (see Figure 5.6 for the output):

```
---
title: "Text Annotations"
output:
  flexdashboard::flex_dashboard:
    orientation: rows
---


Monthly deaths from bronchitis, emphysema and asthma in the
UK, 1974-1979 (Source: P. J. Diggle, 1990, Time Series: A
Biostatistical Introduction. Oxford, table A.3)

```{r setup, include=FALSE}
library(dygraphs)
```
```

```
` ` `

Row {data-height=600}
-------------------------------------

### All Lung Deaths

` ` `{r}
dygraph(ldeaths)
` ` `

Row {data-height=400}
-------------------------------------

### Male Deaths

` ` `{r}
dygraph(mdeaths)
` ` `

> Monthly deaths from lung disease in the UK, 1974-1979

### About dygraphs

This example makes use of the dygraphs R package. The dygraphs
package provides rich facilities for charting time-series data
in R. You can use dygraphs at the R console, within R Markdown
documents, and within Shiny applications.
```

Each component within a dashboard includes optional title and notes sections. The title is simply the text after the third-level (###) section heading. The notes are any text prefaced with > after the code chunk that yields the component's output (see the second component of the above example).

You can exclude the title entirely by applying the .no-title attribute to a section heading.

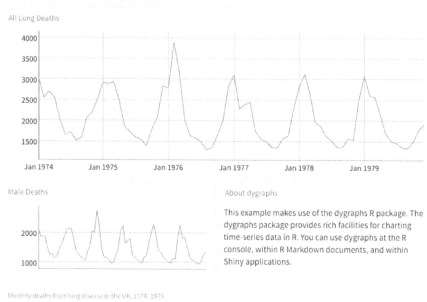

FIGURE 5.6: Text annotations on a dashboard.

5.2.4   Navigation bar

By default, the dashboard navigation bar includes the document's title, au-thor, and date. When a dashboard has multiple pages (Section 5.1.3), links to the various pages are also included on the left side of the navigation bar. You can also add social links to the dashboard.

In addition, you can add custom links to the navigation bar using the navbar option. For example, the following options add an "About" link to the navi-gation bar:

```
---
title: "Navigation Bar"
output:
  flexdashboard::flex_dashboard:
    navbar:
```

```
      - { title: "About", href: "https://example.com/about" }
---
```

Navigation bar items must include either a `title` or `icon` field (or both). You should also include a `href` as the navigation target. The `align` field is optional (it can be "left" or "right" and defaults to "right").

You can include links to social sharing services via the `social` option. For example, the following dashboard includes Twitter and Facebook links as well as a drop-down menu with a more complete list of services:

```
---
title: "Social Links"
output:
  flexdashboard::flex_dashboard:
    social: [ "twitter", "facebook", "menu" ]
---
```

The `social` option can include any number of the following services: `"facebook"`, `"twitter"`, `"google-plus"`, `"linkedin"`, and `"pinterest"`. You can also specify `"menu"` to provide a generic sharing drop-down menu that includes all of the services.

5.3 Shiny

By adding Shiny to a dashboard, you can let viewers change underlying parameters and see the results immediately, or let dashboards update themselves incrementally as their underlying data changes (see functions `reactiveFileReader()` and `reactivePoll()` in the **shiny** package). This is done by adding `runtime: shiny` to a standard dashboard document, and then adding one or more input controls and/or reactive expressions that dynamically drive the appearance of the components within the dashboard.

Using Shiny with **flexdashboard** turns a static R Markdown report into an interactive document. It is important to note that interactive documents need to be deployed to a Shiny Server to be shared broadly (whereas static R Markdown documents are standalone web pages that can be attached to emails or served from any standard web server).

Note that the **shinydashboard**[1] package provides another way to create dashboards with Shiny.

5.3.1 Getting started

The steps required to add Shiny components to a dashboard are:

1. Add `runtime: shiny` to the options declared at the top of the document (YAML metadata).

2. Add the `{.sidebar}` attribute to the first column of the dashboard to make it a host for Shiny input controls (note that this step is not strictly required, but this will generate a typical layout for Shiny-based dashboards).

3. Add Shiny inputs and outputs as appropriate.

4. When including plots, be sure to wrap them in a call to `render-Plot()`. This is important not only for dynamically responding to changes, but also to ensure that they are automatically re-sized when their container changes.

5.3.2 A Shiny dashboard example

Here is a simple example of a dashboard that uses Shiny (see Figure 5.7 for the output):

```
---
title: "Old Faithful Eruptions"
output: flexdashboard::flex_dashboard
runtime: shiny
---

```{r global, include=FALSE}
# load data in 'global' chunk so it can be shared
```

---

[1]https://rstudio.github.io/shinydashboard/
```

```
# by all users of the dashboard
library(datasets)
data(faithful)
```
```

Column {.sidebar}
--------------------------------------------------

Waiting time between eruptions and the duration of the eruption
for the Old Faithful geyser in Yellowstone National Park,
Wyoming, USA.

```{r}
selectInput(
  "n_breaks", label = "Number of bins:",
  choices = c(10, 20, 35, 50), selected = 20
)

sliderInput(
  "bw_adjust", label = "Bandwidth adjustment:",
  min = 0.2, max = 2, value = 1, step = 0.2
)
```
```

Column

Geyser Eruption Duration

```{r}
renderPlot({
  erpt = faithful$eruptions
  hist(
    erpt, probability = TRUE, breaks = as.integer(input$n_breaks),
    xlab = "Duration (minutes)", main = "Geyser Eruption Duration",
    col = 'gray', border = 'white'
  )
```

```
dens = density(erpt, adjust = input$bw_adjust)
lines(dens, col = "blue", lwd = 2)
})
```
```

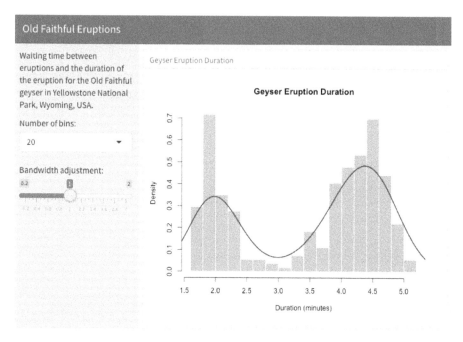

**FIGURE 5.7:** An interactive dashboard based on Shiny.

The first column includes the {.sidebar} attribute and two Shiny input controls; the second column includes the Shiny code required to render the chart based on the inputs.

One important thing to note about this example is the chunk labeled global at the top of the document. The global chunk has special behavior within **flexdashboard**: it is executed only once within the global environment, so that its results (e.g., data frames read from disk) can be accessed by all users of a multi-user dashboard. Loading your data within a global chunk will result in substantially better startup performance for your users, and hence is highly recommended.
```

5.3.3 Input sidebar

You add an input sidebar to a flexdashboard by adding the {.sidebar} attribute to a column, which indicates that it should be laid out flush to the left with a default width of 250 pixels and a special background color. Sidebars always appear on the left no matter where they are defined within the flow of the document.

If you are creating a dashboard with multiple pages, you may want to use a single sidebar that applies across all pages. In this case, you should define the sidebar using a *first-level* Markdown header.

5.3.4 Learning more

Below are some good resources for learning more about Shiny and creating interactive documents:

1. The official Shiny website (`http://shiny.rstudio.com`) includes extensive articles, tutorials, and examples to help you learn more about Shiny.

2. The article "Introduction to Interactive Documents[2]" on the Shiny website is a great guide for getting started with Shiny and R Markdown.

3. For deploying interactive documents, you may consider Shiny Server or RStudio Connect: `https://www.rstudio.com/products/shiny/shiny-server/`.

---

[2]`http://shiny.rstudio.com/articles/interactive-docs.html`

6

Tufte Handouts

The Tufte handout style is a style that Edward Tufte[1] uses in his books and handouts. Tufte's style is known for its extensive use of sidenotes, tight integration of graphics with text, and well-set typography. This style has been implemented in LaTeX and HTML/CSS,[2] respectively. Both implementations have been ported into the **tufte** package (Xie and Allaire, 2018). If you want LaTeX/PDF output, you may use the `tufte_handout` format for handouts, and `tufte_book` for books. For HTML output, use `tufte_html`, e.g.,

```
---
title: "An Example Using the Tufte Style"
author: "John Smith"
output:
  tufte::tufte_handout: default
  tufte::tufte_html: default
---
```

Figure 6.1 shows the basic layout of the Tufte style, in which you can see a main column on the left that contains the body of the document, and a side column on the right to display sidenotes.

There are two goals for the **tufte** package:

1. To produce both PDF and HTML output with similar styles from the same R Markdown document.

2. To provide simple syntax to write elements of the Tufte style such as side notes and margin figures. For example, when you want a margin figure, all you need to do is the chunk option `fig.margin` = `TRUE`, and **tufte** will take care of the details for

---

[1] https://en.wikipedia.org/wiki/Edward_Tufte
[2] See Github repositories https://github.com/tufte-latex/tufte-latex and https://github.com/edwardtufte/tufte-css.

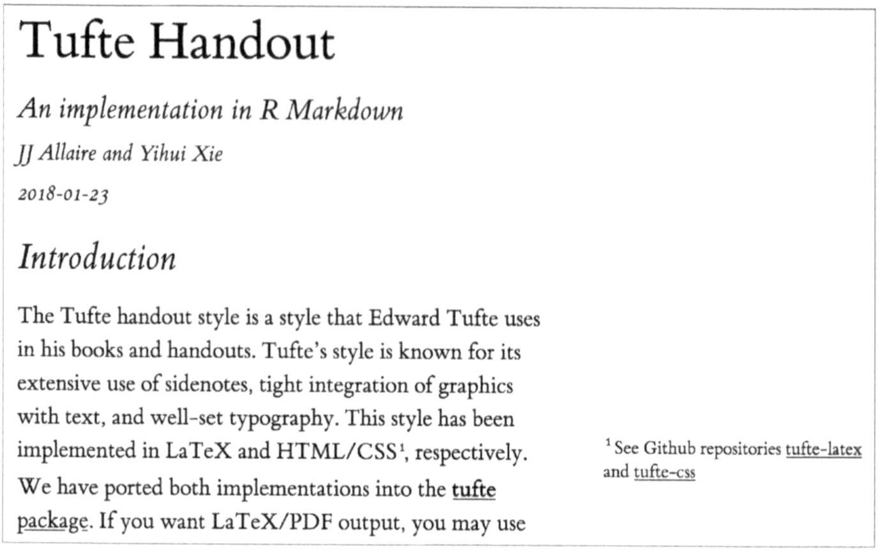

Tufte Handout

An implementation in R Markdown

JJ Allaire and Yihui Xie

2018-01-23

Introduction

The Tufte handout style is a style that Edward Tufte uses in his books and handouts. Tufte's style is known for its extensive use of sidenotes, tight integration of graphics with text, and well-set typography. This style has been implemented in LaTeX and HTML/CSS[1], respectively. We have ported both implementations into the **tufte** package. If you want LaTeX/PDF output, you may use

[1] See Github repositories tufte-latex and tufte-css

FIGURE 6.1: The basic layout of the Tufte style.

> you, so you never need to think about LaTeX environments like \begin{marginfigure} \end{marginfigure} or HTML tags like <span class="marginfigure"> </span>; the LaTeX and HTML code under the hood may be complicated, but you never need to learn or write such code.

You can use the wizard in RStudio IDE from the menu File -> New File -> R Markdown -> From Template to create a new R Markdown document with a default example provided by the **tufte** package. Note that you need a LaTeX distribution if you want PDF output (see Chapter 1).

6.1 Headings

The Tufte style provides the first and second-level headings (that is, # and ##), demonstrated in the next section. You may get unexpected output (and even errors) if you try to use ### and smaller headings.

In his later books,[3] Tufte starts each section with a bit of vertical space, a non-indented paragraph, and sets the first few words of the sentence in small caps. To accomplish this using this style, call the `newthought()` function in **tufte** in an *inline R expression* `` `r ` ``. Note that you should not assume **tufte** has been attached to your R session. You should either use `library(tufte)` in your R Markdown document before you call `newthought()`, or use `tufte::newthought()`.

6.2 Figures

6.2.1 Margin figures

Images and graphics play an integral role in Tufte's work. To place figures in the margin, you can use the **knitr** chunk option `fig.margin = TRUE`. For example:

```
```{r fig-margin, fig.margin=TRUE}
plot(cars)
```
```

As in other Rmd documents, you can use the `fig.cap` chunk option to provide a figure caption, and adjust figure sizes using the `fig.width` and `fig.height` chunk options, which are specified in inches, and will be automatically scaled down to fit within the handout margin.

Figure 6.2 shows what a margin figure looks like.

6.2.2 Arbitrary margin content

You can include anything in the margin using the **knitr** engine named `marginfigure`. Unlike R code chunks `` ```{r} ``, you write a chunk starting with `` ```{marginfigure} `` instead, then put the content in the chunk, e.g.,

---

[3]Such as "Beautiful Evidence": `http://www.edwardtufte.com/tufte/books_be`.

Images and graphics play an integral role in Tufte's work.
To place figures in the margin you can use the **knitr**
chunk option `fig.margin` = `TRUE`. For example:

```
library(ggplot2)
mtcars2 <- mtcars
mtcars2$am <- factor(
  mtcars$am, labels = c('automatic', 'manual')
)
ggplot(mtcars2, aes(hp, mpg, color = am)) +
  geom_point() + geom_smooth() +
  theme(legend.position = 'bottom')
```

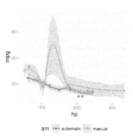

MPG vs horsepower, colored by transmission.

Note the use of the `fig.cap` chunk option to provide a
figure caption. You can adjust the proportions of figures
using the `fig.width` and `fig.height` chunk options.

FIGURE 6.2: A margin figure in the Tufte style.

````
```{marginfigure}
We know from _the first fundamental theorem of calculus_ that
for $x$ in $[a, b]$:
$$\frac{d}{dx}\left( \int_{a}^{x} f(u)\,du\right)=f(x).$$
```
````

For the sake of portability between LaTeX and HTML, you should keep
the margin content as simple as possible (syntax-wise) in the `marginfigure`
blocks. You may use simple Markdown syntax like `**bold**` and `_italic_`
text, but please refrain from using footnotes, citations, or block-level elements
(e.g., blockquotes and lists) there.

Note that if you set `echo` = `FALSE` in your global chunk options, you will
have to add `echo` = `TRUE` to the chunk to display a margin figure, for example
` ```{marginfigure, echo = TRUE}`.

6.2.3 Full-width figures

You can arrange for figures to span across the entire page by using the chunk
option `fig.fullwidth` = `TRUE`, e.g.,

```{r, fig.width=10, fig.height=2, fig.fullwidth=TRUE}
par(mar = c(4, 4, .1, .2)); plot(sunspots)
```

Other chunk options related to figures can still be used, such as `fig.width`, `fig.cap`, and `out.width`, etc. For full-width figures, usually `fig.width` is large and `fig.height` is small. In the above example, the plot size is 10x2.

Figure 6.3 shows what a full-width figure looks like.

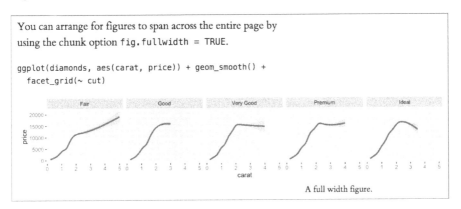

You can arrange for figures to span across the entire page by using the chunk option `fig.fullwidth = TRUE`.

```
ggplot(diamonds, aes(carat, price)) + geom_smooth() +
  facet_grid(~ cut)
```

A full width figure.

FIGURE 6.3: A full-width figure in the Tufte style.

6.2.4 Main column figures

Besides margin and full-width figures, you can certainly also include figures constrained to the main column. This is the default type of figures in the LaTeX/HTML output, and requires no special chunk options.

Figure 6.4 shows what a figure looks like in the main column.

6.3 Sidenotes

One of the most prominent and distinctive features of this style is the extensive use of sidenotes. There is a wide margin to provide ample room for sidenotes and small figures. Any use of a footnote, of which the Markdown

Besides margin and full width figures, you can of course
also include figures constrained to the main column.
This is the default type of figures in the LaTeX/HTML
output.

```
ggplot(diamonds, aes(cut, price)) + geom_boxplot()
```

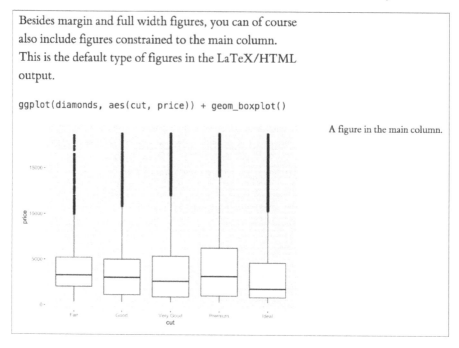

A figure in the main column.

FIGURE 6.4: A figure in the main column in the Tufte style.

syntax is ^[footnote content], will automatically be converted to a sidenote.

If you would like to place ancillary information in the margin without the
sidenote mark (the superscript number), you can use the margin_note()
function from **tufte** in an inline R expression. This function does not process
the text with Pandoc, so Markdown syntax will not work here. If you need
to write anything in Markdown syntax, please use the marginfigure block
described previously.

6.4   References

References can be displayed as margin notes for HTML output. To enable this
feature, you must set link-citations: yes in the YAML metadata, and the
version of pandoc-citeproc should be at least 0.7.2. To check the version of
pandoc-citeproc in your system, you may run this in R:

```
system2("pandoc-citeproc", "--version")
```

If your version of pandoc-citeproc is too low, or you did not set link-citations: yes in YAML, references in the HTML output will be placed at the end of the output document.

You can also explicitly disable this feature via the margin_references option, e.g.,

```
---
output:
  tufte::tufte_html:
    margin_references: false
---
```

6.5 Tables

You can use the kable() function from the **knitr** package to format tables that integrate well with the rest of the Tufte handout style. The table captions are placed in the margin like figures in the HTML output. A simple example (Figure 6.5 shows the output):

````
```{r}
knitr::kable(
  mtcars[1:6, 1:6], caption = 'A subset of mtcars.'
)
```
````

6.6 Block quotes

We know from the Markdown syntax that paragraphs that start with > are converted to block quotes. If you want to add a right-aligned footer for the

```
knitr::kable(
  mtcars[1:6, 1:6], caption = 'A subset of mtcars.'
)
```

	mpg	cyl	disp	hp	drat	wt	A subset of mtcars.
Mazda RX4	21.0	6	160	110	3.90	2.620	
Mazda RX4 Wag	21.0	6	160	110	3.90	2.875	
Datsun 710	22.8	4	108	93	3.85	2.320	
Hornet 4 Drive	21.4	6	258	110	3.08	3.215	
Hornet Sportabout	18.7	8	360	175	3.15	3.440	
Valiant	18.1	6	225	105	2.76	3.460	

FIGURE 6.5: A table in the Tufte style.

quote, you may use the function `quote_footer()` from **tufte** in an inline R expression. Here is an example:

```
> "If it weren't for my lawyer, I'd still be in prison.
>  It went a lot faster with two people digging."
>
> `r tufte::quote_footer('--- Joe Martin')`
```

6.7 Responsiveness

The HTML page is responsive in the sense that when the page width is smaller than 760px, sidenotes and margin notes will be hidden by default. For sidenotes, you can click their numbers (the superscripts) to toggle their visibility. For margin notes, you may click the circled plus signs to toggle visibility (see Figure 6.6).

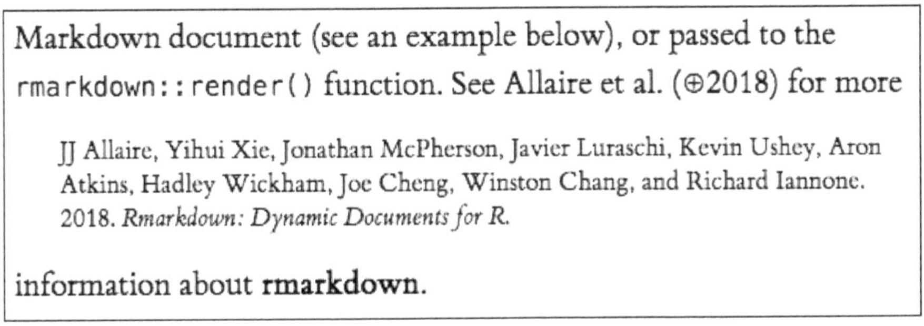

Markdown document (see an example below), or passed to the `rmarkdown::render()` function. See Allaire et al. (⊕2018) for more

> JJ Allaire, Yihui Xie, Jonathan McPherson, Javier Luraschi, Kevin Ushey, Aron Atkins, Hadley Wickham, Joe Cheng, Winston Chang, and Richard Iannone. 2018. *Rmarkdown: Dynamic Documents for R.*

information about **rmarkdown.**

FIGURE 6.6: The Tufte HTML style on narrow screens.

6.8 Sans-serif fonts and epigraphs

There are a few other things in Tufte CSS that we have not mentioned so far. If you prefer sans-serif fonts, use the function `sans_serif()` in **tufte**. For epigraphs, you may use a pair of underscores to make the paragraph italic in a block quote, e.g.,

```
> _I can win an argument on any topic, against any opponent.
> People know this, and steer clear of me at parties. Often,
> as a sign of their great respect, they don't even invite me._
>
> `r quote_footer('--- Dave Barry')`
```

6.9 Customize CSS styles

You can turn on/off some features of the Tufte style in HTML output. The default features enabled are:

```
---
output:
  tufte::tufte_html:
```

```
    tufte_features: ["fonts", "background", "italics"]
---
```

If you do not want the page background to be lightyellow, you can remove background from `tufte_features`. You can also customize the style of the HTML page via a CSS file. For example, if you do not want the subtitle to be italic, you can define

```
h3.subtitle em {
  font-style: normal;
}
```

in, say, a CSS file `my-style.css` (under the same directory of your Rmd document), and apply it to your HTML output via the `css` option, e.g.,

```
---
output:
  tufte::tufte_html:
    tufte_features: ["fonts", "background"]
    css: "my-style.css"
---
```

There is also a variant of the Tufte style in HTML/CSS named "Envisioned CSS[4]". This style can be enabled by specifying the argument `tufte_variant = 'envisioned'` in `tufte_html()`,[5] e.g.,

```
---
output:
  tufte::tufte_html:
    tufte_variant: "envisioned"
---
```

You can see a live example at `https://rstudio.github.io/tufte/`. It is also available in Simplified Chinese: `https://rstudio.github.io/tufte/cn/`, and its envisioned style can be found at `https://rstudio.github.io/tufte/envisioned/`.

---

[4]`http://nogginfuel.com/envisioned-css/`
[5]The actual Envisioned CSS was not used in the **tufte** package. Only the fonts, background color, and text color are changed based on the default Tufte style.

7

xaringan Presentations

We have introduced a few HTML5 presentation formats in Chapter 4. The **xaringan** package (Xie, 2018g) is an R Markdown extension based on the JavaScript library remark.js (`https://remarkjs.com`) to generate HTML5 presentations of a different style. See Figure 7.1 for two sample slides.

FIGURE 7.1: Two sample slides created from the xaringan package.

The name "xaringan" came from Sharingan (`http://naruto.wikia.com/wiki/Sharingan`) in the Japanese manga and anime "Naruto". The word was deliberately chosen to be difficult to pronounce for most people (unless you have watched the anime), because its author (me) loved the style very much, and was concerned that it would become too popular.[1] The concern was somewhat naive, because the style is actually very customizable, and users started to contribute more themes to the package later.

The **xaringan** package is based on the JavaScript library remark.js (`https://remarkjs.com`); remark.js only supports Markdown, and **xaringan** added the support for R Markdown as well as other utilities to make it easier to build and preview slides.

You can learn more about the background stories and the usage of the

---

[1] The main reason I stopped using LaTeX Beamer slides was because of its popularity: when you attend academic conferences, you see Beamer slides everywhere.

xaringan package from the documentation at `http://slides.yihui.name/` `xaringan/`, which is actually a set of slides generated from **xaringan**. You may also read a potentially biased blog post of mine to know why I preferred **xaringan** / remark.js for HTML5 presentations: `https://yihui.name/en/` `2017/08/why-xaringan-remark-js/`.

7.1   Get started

You can install either the CRAN version or the development version on GitHub (`https://github.com/yihui/xaringan`):

```
# install from CRAN
install.packages('xaringan')

# or GitHub
devtools::install_github('yihui/xaringan')
```

If you use RStudio, it is easy to get started from the menu `File -> New File -> R Markdown -> From Template -> Ninja Presentation`, and you will see an R Markdown example in the editor. Press the `Knit` button to compile it, or use the RStudio addin `Infinite Moon Reader` to live preview the slides: every time you update and save the Rmd document, the slides will be automatically reloaded.

The main R Markdown output format in this package is `moon_reader()`. See the R help page `?xaringan::moon_reader` for all possible configurations. Below is a quick example:

```
---
title: "Presentation Ninja"
subtitle: "with xaringan"
author: "Yihui Xie"
date: "2016/12/12"
output:
  xaringan::moon_reader:
    lib_dir: libs
```

```
    nature:
      highlightStyle: github
      countIncrementalSlides: false
---


One slide.


---


Another slide.
```

7.2 Keyboard shortcuts

After opening slides generated from **xaringan** or remark.js, you may press the key h (Help) or ? on your keyboard to learn all possible keyboard shortcuts, which may help you better present your slides.

- To go the previous slide, you may press Up/Left arrows, PageUp, or k.

- To go the next slide, you may press Right/Down arrows, PageDown, Space, or j.

- You may press Home to go to the first slide, or End to go to the last slide, if you have these keys.

- Typing a number and pressing Return (or Enter), you can jump to a specific slide with that page number.

- Press b to black out a slide, and m to "mirror" a slide (reverse everything on the slide). These techniques can be useful when you do not want the audience to read the slide, e.g., when you have solutions on a slide but do not want to show them to your students immediately. I encourage you to try m; it can be a lot of fun. You can press these keys again to resume the normal slide.

- Press f to toggle the fullscreen mode.

- Press c to clone the slides to a new browser window; slides in the two windows will be in sync as you navigate through them. Press p to toggle the presenter mode. The presenter mode shows thumbnails of the current slide and the next slide on the left, presenter notes on the right (see Section 7.3.5), and also a timer on the top right. The keys c and p can be very useful when you present with your own computer connected to a second screen (such as a projector). On the second screen, you can show the normal slides, while cloning the slides to your own computer screen and using the presenter mode. Only you can see the presenter mode, which means only you can see presenter notes and the time, and preview the next slide. You may press t to restart the timer at any time.

- Press h or ? again to exit the help page.

7.3   Slide formatting

The remark.js Wiki[2] contains detailed documentation about how to format slides and use the presentation (keyboard shortcuts). The **xaringan** package has simplified several things compared to the official remark.js guide, e.g., you do not need a boilerplate HTML file, you can set the autoplay mode via an option of `moon_reader()`, and LaTeX math basically just works.

Please note that remark.js has its own Markdown interpreter that is *not compatible* with Pandoc's Markdown converter, so you will not be able to use any advanced Pandoc Markdown features (e.g., the citation syntax `[@key]`). You may use raw HTML when there is something you desire that is not supported by remark.js. For example, you can generate an HTML table via `knitr::kable(head(iris), 'html')`.

7.3.1   Slides and properties

Every new slide is created under a horizontal rule (`---`). The content of the slide can be arbitrary, e.g., it does not have to have a slide title, and if it does, the title can be of any level you prefer (`#`, `##`, or `###`).

A slide can have a few properties, including `class` and `background-image`, etc. Properties are written in the beginning of a slide, e.g.,

---

[2]`https://github.com/gnab/remark/wiki`

```
---

class: center, inverse
background-image: url("images/cool.png")

# A new slide

Content.
```

The `class` property assigns class names to the HTML tag of the slide, so that you can use CSS to style specific slides. For example, for a slide with the `inverse` class, you may define the CSS rules (to render text in white on a dark background):

```
.inverse {
  background-color: #272822;
  color: #d6d6d6;
  text-shadow: 0 0 20px #333;
}
```

Then include the CSS file (say, `my-style.css`) via the `css` option of `xaringan::moon_reader`:

```
---

output:
  xaringan::moon_reader:
    css: "my-style.css"

---
```

Actually the style for the `inverse` class has been defined in the default theme of **xaringan**, so you do not really need to define it again unless you want to override it.

Other available class names are `left`, `center`, and `right` for the horizontal alignment of all elements on a slide, and `top`, `middle`, and `bottom` for the vertical alignment.

Background images can be set via the `background-image` property. The image can be either a local file or an online image. The path should be put inside

url(), which is the CSS syntax. You can also set the background image size
and position, e.g.,

```
background-image: url("`r xaringan:::karl`")
background-position: center
background-size: contain
```

All these properties require you to understand CSS.[3] In the above example,
we actually used an inline expression xaringan::karl to return a URL of an
image of Karl Broman (http://kbroman.org), which is one of the highlights
of the **xaringan** package.

7.3.2   The title slide

There is a special slide, the title slide, that is automatically generated from the
YAML metadata of your Rmd document. It contains the title, subtitle, author,
and date (all are optional). This slide has the classes inverse, center, middle,
and title-slide by default, which looks like the left image in Figure 7.1. If
you do not like the default style, you may either customize the .title-slide
class, or provide a custom vector of classes via the titleSlideClass option
under the nature option, e.g.,

```
---
output:
  xaringan::moon_reader:
    nature:
      titleSlideClass: ["right", "top", "my-title"]
---
```

You can also disable the automatic title slide via the seal option and create
one manually by yourself:

```
---
output:
  xaringan::moon_reader:
    seal: false
```

---

[3]There are many tutorials online if you search for "CSS background", e.g., https://www.
w3schools.com/cssref/css3_pr_background.asp.

```
---

# My Own Title

### Author

Whatever you want to put on the title slide.
```

7.3.3   Content classes

You can assign classes to any elements on a slide, too. The syntax is `.class-Name[content]`. This is a very powerful feature of remark.js, and one of very few features not available in Pandoc. Basically it makes it possible to style any elements on a slide via CSS. There are a few built-in content classes, `.left[ ]`, `.center[ ]`, and `.right[ ]`, to align elements horizontally on a slide, e.g., you may center an image:

```
.center[![description of the image](images/foo.png)]
```

The content inside `[ ]` can be anything, such as several paragraphs, or lists. The default theme of **xaringan** has provided four more content classes:

- `.left-column[ ]` and `.right-column[ ]` provide a sidebar layout. The left sidebar is narrow (20% of the slide width), and the right column is the main column (75% of the slide width). If you have multiple level-2 (`##`) or level-3 (`###`) headings in the left column, the last heading will be highlighted, with previous headings being grayed out.

- `.pull-left[ ]` and `.pull-right[ ]` provide a two-column layout, and the two columns are of the same width. Below is an example:

```
.pull-left[
- One bullet.

- Another bullet.
]

.pull-right[
```

```
![an image](foo.png)
]
```

You can design your own content classes if you know CSS, e.g., if you want to make text red via `.red[ ]`, you may define this in CSS:

```
.red { color: red; }
```

7.3.4   Incremental slides

When you want to show content incrementally on a slide (e.g., holding a funny picture until the last moment), you can use two dashes to separate the content. The two dashes can appear anywhere except inside content classes, so you can basically split your content in any way you like, e.g.,

```
---

# Two dashes

The easiest way to build incremental slides is...
--

  to use two dashes `--` to separate content on a slide.

--

You can divide a slide in _any way you want_.

--

- One bullet

- Another bullet

--

- And one more
```

```
--

.center[
![Saw](https://slides.yihui.name/gif/saw-branch.gif)

Don't saw your slides too hard.
]
```

There are a few other advanced ways to build incremental slides documented in the presentation at `https://slides.yihui.name/xaringan/incremental.html`.

7.3.5   Presenter notes

You can write notes for yourself to read in the presenter mode (press the keyboard shortcut p). These notes are written under three question marks ??? after a slide, and the syntax is also Markdown, which means you can write any elements supported by Markdown, such as paragraphs, lists, images, and so on. For example:

```
---

The holy passion of Friendship is of so sweet and steady
and loyal and enduring a nature that it will last through
a whole lifetime...

???

_if not asked to lend money_.

--- Mark Twain
```

A common mistake in presentations, especially for presenters without much experience, is to stuff a slide with too much content. The consequence is either a speaker, out of breath, reading the so many words out loud, or the audience starting to read the slides quietly by themselves without listening. Slides are not papers or books, so you should try to be brief in the visual content of slides but verbose in verbal narratives. If you have a lot to say about a slide, but cannot remember everything, you may consider using presenter notes.

I want to mention a technical note about the presenter mode: when connecting to a projector, you should make sure not to mirror the two screens. Instead, separate the two displays, so you can drag the window with the normal view of slides to the second screen. Figure 7.2 shows how to do it from the "System Preferences" on macOS (do not check the box "Mirror Displays").

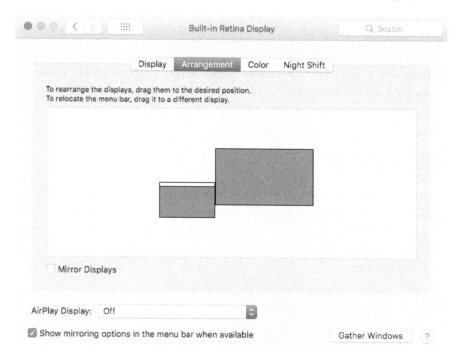

FIGURE 7.2: Separate the current display from the external display.

7.3.6 yolo: true

Inspired by a random feature request from a tweet by Karthik Ram[4], the output format xaringan::moon_reader provided an option named yolo (an acronym of "you only live once"). If you set it to true, a photo of Karl Broman (with a mustache) will be inserted into a random slide in your presentation.[5]

---

[4]https://twitter.com/_inundata/status/798970002992873472
[5]For the full story behind the mustache, see Karl's post at http://kbroman.org/blog/2014/08/28/the-mustache-photo/.

```
---
output:
  xaringan::moon_reader:
    yolo: true
---
```

The **xaringan** package is probably best known for this feature. I want to thank Karl for letting me use this photo. It always makes me happy for mysterious reasons.

A less well-known feature is that you can actually replace Karl's picture with other pictures, and/or specify how many times you want a picture to randomly show up in your presentation. For example:

```
---
output:
  xaringan::moon_reader:
    yolo:
      img: kangaroo.jpg
      times: 5
---
```

Developing software is fun, isn't it?

7.4 Build and preview slides

You can knit the source document like other Rmd documents to view the output, but it may be tedious to have to knit it over and over again whenever you make changes. The other way to preview the slides is the RStudio addin "Infinite Moon Reader" or the function xaringan::inf_mr(), as mentioned in Section 2.2. With this way, you can continuously preview your slides just by saving the source document. The continuous preview is achieved via a local web server launched by the **servr** package (Xie, 2018e).

One distinction of xaringan::moon_reader when compared to other R Markdown output formats is that it does not generate self-contained HTML documents by default (see Section 3.1.9). This means none of the external dependencies such as images or JavaScript libraries will be embedded in the HTML output file by default. Due to technical difficulties (remark.js does not use Pandoc but renders Markdown in real time in the browser), it is hard to implement the self-contained mode well. If you have to publish the slides to a web server, but it is not convenient for you to upload all the dependencies, **xaringan** may not be a good choice for you. If you use GitHub Pages or Netlify, this may not be a big problem (you commit or upload all files).

7.5 CSS and themes

The format `xaringan::moon_reader` has a `css` option, to which you can pass a vector of CSS file paths, e.g.,

```
---
output:
  xaringan::moon_reader:
    css: ["default", "extra.css"]
---
```

In theory, the file path should contain the extension `.css`. If a path does not contain a filename extension, it is assumed to be a built-in CSS file in the **xaringan** package. For example, `default` in the above example means `default.css` in the package under the path `xaringan:::pkg_resource()`. To see all built-in CSS files, call `xaringan:::list_css()` in R.

When you only want to override a few CSS rules in the default theme, you do not have to copy the whole file `default.css`; instead, create a new (and hopefully smaller) CSS file that only provides new CSS rules.

Users have contributed a few themes to **xaringan**. For example, you can use the `metropolis` theme (`https://github.com/pat-s/xaringan-metropolis`):

```
---
output:
  xaringan::moon_reader:
```

```
    css: [default, metropolis, metropolis-fonts]
---
```

To see all possible themes:

```
names(xaringan:::list_css())
```

```
##  [1] "default-fonts"    "default"
##  [3] "duke-blue"        "hygge-duke"
##  [5] "hygge"            "metropolis-fonts"
##  [7] "metropolis"       "rladies-fonts"
##  [9] "rladies"          "robot-fonts"
## [11] "robot"            "rutgers-fonts"
## [13] "rutgers"          "tamu-fonts"
## [15] "tamu"             "uo-fonts"
## [17] "uo"
```

If you also want to contribute themes, please read the guide at `https://yihui.name/en/2017/10/xaringan-themes`.

7.6 Some tips

Lastly, we present a few tips that may help you make better presentations.

7.6.1 Autoplay slides

Slides can be automatically played if you set the `autoplay` option under `nature` (in milliseconds). For example, the next slide can be displayed automatically every 30 seconds in a lightning talk:

```
---
output:
  xaringan::moon_reader:
    nature:
```

```
    autoplay: 30000
---
```

7.6.2   Countdown timer

A countdown timer can be added to every page of the slides using the count-down option under `nature`. For example, if you want to spend one minute on every page when you give the talk, you can set:

```
---
output:
  xaringan::moon_reader:
    nature:
      countdown: 60000
---
```

Then you will see a timer counting down from `01:00`, to `00:59`, `00:58`, ... When the time is out, the timer will continue but the time turns red.

7.6.3   Highlight code lines

The option `highlightLines: true` of `nature` will highlight code lines that start with `*`, or are wrapped in `{{ }}`, or have trailing comments `#<<`:

```
---
output:
  xaringan::moon_reader:
    nature:
      highlightLines: true
---
```

Below are a few examples:

```r
if (TRUE) {
* message("Very important!")
}
```

```
```
```

```{r tidy=FALSE}
if (TRUE) {
{{ message("Very important!") }}
}
```

```{r tidy=FALSE}
library(ggplot2)
ggplot(mtcars) +
  aes(mpg, disp) +
  geom_point() +    #<<
  geom_smooth()     #<<
```

Note that the first way does not give you valid R code in the source document, but the latter two ways provide syntactically valid R code, and in the output slides, you will not see the tokens {{ }} or #<<. The lines will be highlighted with a yellow background by default.

7.6.4 Working offline

To make slides work offline, you need to download a copy of remark.js in advance, because **xaringan** uses the online version by default. You can use `xaringan::summon_remark()` to download the latest or a specified version of remark.js. By default, it is downloaded to `libs/remark-latest.min.js`.

Then change the `chakra` option in the YAML metadata to point to this file, e.g.,

```
output:
  xaringan::moon_reader:
    chakra: libs/remark-latest.min.js
```

Making the slides work offline can be tricky, since you may have other dependencies. The remark.js dependency is easy to deal with because it is a single JavaScript file; other dependencies such as MathJax can be extremely tricky. If you used Google web fonts in slides (the default theme uses *Yanone Kaf-*
```

feesatz, Droid Serif, and *Source Code Pro*), they will not work offline unless you download or install them locally. The Heroku app google-webfonts-helper[6] can help you download fonts and generate the necessary CSS.

7.6.5 Macros

The Markdown syntax of remark.js can be amazingly extensible, because it allows users to define custom macros (JavaScript functions) that can be applied to Markdown text using the syntax `![:macroName arg1, arg2, ...]` or `![:macroName arg1, arg2](this)`. For example, you can define a macro named `scale` to set the width of an image:

```
remark.macros.scale = function(w) {
  var url = this;
  return '<img src="' + url + '" style="width: ' + w + '" />';
};
```

Then the Markdown text

```
![:scale 50%](image.jpg)
```

will be translated to:

```
<img src="image.jpg" style="width: 50%" />
```

Now you should see that you can use cleaner pseudo-Markdown syntax to generate HTML.

To insert macros in **xaringan** slides, you can save your macros in a file (e.g., `macros.js`), and use the option `beforeInit` under the option `nature`, e.g.,

```
output:
  xaringan::moon_reader:
    nature:
      beforeInit: "macros.js"
```

The `beforeInit` option can be used to insert arbitrary JavaScript code before

---

[6]https://google-webfonts-helper.herokuapp.com/fonts

remark.js initializes the slides. Inserting macros is just one of its possible applications. For example, when you embed tweets from Twitter in slides, usually you need to load `https://platform.twitter.com/widgets.js`, which can be loaded via the `beforeInit` option.

7.6.6 Disadvantages

The **xaringan** package was originally designed for "ninja", meaning that if you know CSS, you will be able to freely customize the style, otherwise you can only accept the default themes. Playing with CSS can be fun and rewarding, but it can also easily waste your time. You aesthetic standards and taste may change from time to time, and you could end up tweaking the styles all the time.

The HTML output file generated from **xaringan** is not self-contained by default, as we mentioned in Section 7.4. If your slides must be self-contained and cannot be served through a web server, **xaringan** may not be a good option for you.

HTML widgets may not work well in **xaringan**. This might be improved in the future, but it is a little tricky technically.

When printing the slides to PDF from Google Chrome (see Section 4.1.10), I recommend that you open the slides and go through all pages at least once, to make sure all content has been rendered in the browser. Without navigating through all slides manually once, some content may not be printed correctly (such as MathJax expressions and HTML widgets).

8

reveal.js Presentations

The **revealjs** package (El Hattab and Allaire, 2017) provides an output format `revealjs::revealjs_presentation` that can be used to create yet another style of HTML5 slides based on the JavaScript library reveal.js[1]. You may install the R package from CRAN:

```r
install.packages("revealjs")
```

To create a reveal.js presentation from R Markdown, you specify the `revealjs_presentation` output format in the YAML metadata of your document. You can create a slide show broken up into sections by using the # and ## heading tags; you can also create a new slide without a header using a horizontal rule (---). For example, here is a simple slide show:

```yaml
---
title: "Habits"
author: John Doe
date: March 22, 2005
output: revealjs::revealjs_presentation
---
```

```
# In the morning

## Getting up

- Turn off alarm
- Get out of bed

## Breakfast
```

---

[1]https://revealjs.com

```
- Eat eggs
- Drink coffee

# In the evening

## Dinner

- Eat spaghetti
- Drink wine

## Going to sleep

- Get in bed
- Count sheep
```

See Figure 8.1 for two sample slides.

FIGURE 8.1: Two sample slides created from the revealjs package.

8.1   Display modes

The following single character keyboard shortcuts enable alternate display modes:

- 'f': enable fullscreen mode.

- `'o'`: enable overview mode.

Pressing `Esc` exits all of these modes.

___

8.2   Appearance and style

There are several options that control the appearance of reveal.js presentations:

- `theme` specifies the theme to use for the presentation (available themes are `"default"`, `"simple"`, `"sky"`, `"beige"`, `"serif"`, `"solarized"`, `"blood"`, `"moon"`, `"night"`, `"black"`, `"league"`, and `"white"`).

- `highlight` specifies the syntax highlighting style. Supported styles include `"default"`, `"tango"`, `"pygments"`, `"kate"`, `"monochrome"`, `"espresso"`, `"zenburn"`, and `"haddock"`. Pass null to prevent syntax highlighting.

- `center` specifies whether you want to vertically center content on slides (this defaults to `false`).

- `smart` indicates whether to produce typographically correct output, converting straight quotes to curly quotes, `---` to em-dashes, `--` to en-dashes, and `...` to ellipses. Note that `smart` is enabled by default.

For example:

```
---
output:
  revealjs::revealjs_presentation:
    theme: sky
    highlight: pygments
    center: true
---
```

8.2.1   Smaller text

If you need smaller text for certain paragraphs, you can enclose text in the `<small>` tag. For example:

```
<small>This sentence will appear smaller.</small>
```

8.3   Slide transitions

You can use the `transition` and `background_transition` options to specify the global default slide transition style:

- `transition` specifies the visual effect when moving between slides. Available transitions are `"default"`, `"fade"`, `"slide"`, `"convex"`, `"concave"`, `"zoom"` or `"none"`.

- `background_transition` specifies the background transition effect when moving between full page slides. Available transitions are `"default"`, `"fade"`, `"slide"`, `"convex"`, `"concave"`, `"zoom"` or `"none"`.

For example:

```
---
output:
  revealjs::revealjs_presentation:
    transition: fade
---
```

You can override the global transition for a specific slide by using the `data-transition` attribute. For example:

```
## Use a zoom transition {data-transition="zoom"}

## Use a faster speed {data-transition-speed="fast"}
```

You can also use different in and out transitions for the same slide. For example:

```
## Fade in, Slide out {data-transition="slide-in fade-out"}

## Slide in, Fade out {data-transition="fade-in slide-out"}
```

8.4   Slide backgrounds

Slides are contained within a limited portion of the screen by default to allow them to fit any display and scale uniformly. You can apply full page backgrounds outside of the slide area by adding a `data-background` attribute to your slide header element. Four different types of backgrounds are supported: `color`, `image`, `video`, and `iframe`. Below are a few examples.

```
## CSS color background {data-background=#ff0000}

## Full size image background {data-background="background.jpeg"}

## Video background {data-background-video="background.mp4"}

## A background page {data-background-iframe="https://example.com"}
```

Backgrounds transition using a `fade` animation by default. This can be changed to a linear sliding transition by specifying the `background-transition:   slide`. Alternatively, you can set `data-background-transition` on any slide with a background to override that specific transition.

8.5   2-D presenations

You can use the `slide_level` option to specify which level of heading will be used to denote individual slides. If `slide_level` is 2 (the default), a two-dimensional layout will be produced, with level-1 headers building horizontally and level-2 headers building vertically. For example:

```
# Horizontal Slide 1

## Vertical Slide 1
```

```
## Vertical Slide 2

# Horizontal Slide 2
```

With this layout, horizontal navigation will proceed directly from "Horizontal Slide 1" to "Horizontal Slide 2", with vertical navigation to "Vertical Slide 1" (and then "Vertical Slide 2", etc.) presented as an option on "Horizontal Slide 1". See Figure 8.1 for an example (note the arrows at the bottom right on the slides).

8.6   Custom CSS

You can add your own CSS to a reveal.js presentation using the `css` option:

```
---
output:
  revealjs::revealjs_presentation:
    css: styles.css
---
```

If you want to override the appearance of particular HTML element document-wide, you need to qualify it with the `.reveal section` preface in your CSS. For example, to change the default text color in paragraphs to blue, you would use:

```
.reveal section p {
  color: blue;
}
```

8.6.1   Slide IDs and classes

You can also target specific slides or classes of slice with custom CSS by adding IDs or classes to the slides headers within your document. For example, the following slide header

```
## Next Steps {#nextsteps .emphasized}
```

would enable you to apply CSS to all of its content using either of the following CSS selectors:

```
#nextsteps {
    color: blue;
}

.emphasized {
    font-size: 1.2em;
}
```

8.6.2 Styling text spans

You can apply classes defined in your CSS file to spans of text by using a span tag. For example:

```
<span class="emphasized">Pay attention to this!</span>
```

8.7 reveal.js options

Reveal.js has many additional options to configure its behavior. You can specify any of these options using `reveal_options`. For example:

```
---
title: "Habits"
output:
  revealjs::revealjs_presentation:
    self_contained: false
    reveal_options:
      slideNumber: true
```

TABLE 8.1: The currently supported reveal.js plugins.

Plugin	Description
notes	Present per-slide notes in a separate browser window.
zoom	Zoom in and out of selected content with Alt+Click.
search	Find a text string anywhere in the slides and show the next occurrence to the user.
chalkboard	Include handwritten notes within a presentation.

```
    previewLinks: true
---
```

You can find documentation on the various available reveal.js options here: `https://github.com/hakimel/reveal.js#configuration`.

8.8   reveal.js plugins

You can enable various reveal.js plugins using the `reveal_plugins` option. Plugins currently supported plugins are listed in Table 8.1.

Note that the use of plugins requires that the `self_contained` option be set to `false`. For example, this presentation includes both the "notes" and "search" plugins:

```
---
title: "Habits"
output:
  revealjs::revealjs_presentation:
    self_contained: false
    reveal_plugins: ["notes", "search"]
---
```

You can specify additional options for the `chalkboard` plugin using reveal_options. For example:

```
---
title: "Habits"
output:
  revealjs::revealjs_presentation:
    self_contained: false
    reveal_plugins: ["chalkboard"]
    reveal_options:
      chalkboard:
        theme: whiteboard
        toggleNotesButton: false
---
```

8.9 Other features

Refer to Section 3.1 for the documentation of other features of reveal.js presentations, including figure options (Section 3.1.5), MathJax equations (Section 3.1.8), keeping Markdown (Section 3.1.10.1), document dependencies (Section 3.1.9), header and before/after body inclusions (Section 3.1.10.2), custom templates (Section 3.1.10.3), Pandoc arguments (Section 3.1.10.5), and shared options (Section 3.1.11). Also see Section 4.1.2 for incremental bullets.

9

Community Formats

Most output formats introduced in this book are created and maintained by the RStudio team. In fact, other members in the R community have also created a number of R Markdown output formats. We mention those formats that we are aware of in this chapter. If you have developed or know other formats, please feel free to suggest that we add them to the page https://rmarkdown.rstudio.com/formats.html.

9.1 Lightweight Pretty HTML Documents

When designing the **rmarkdown** package, we wished it could produce output documents that look pleasant by default, especially for HTML documents. Pandoc does not really style the HTML documents when converting Markdown to HTML, but **rmarkdown** does. As we mentioned in Section 3.1.4, the themes of HTML documents are based on Bootswatch, which actually relies on the Bootstrap library (https://getbootstrap.com). Although these themes look pretty, the major disadvantage is that their file sizes are relatively large. The size of an HTML document created from an empty R Markdown document with the html_document format is about 600Kb, which is roughly the total size of all CSS, JavaScript, and font files in the default theme.

If you are concerned about the file size but still want a fancy theme, you may consider the **prettydoc** package (Qiu, 2018), which has bundled a few pretty themes (yet small in size). This package provides an output format prettydoc::html_pretty. An empty R Markdown document with this format generates an HTML file of about 70Kb.

9.1.1 Usage

The usage of `prettydoc::html_pretty` is very similar to `html_document`, with two major differences:

- The `theme` option takes different values. The currently supported themes are `"cayman"`, `"tactile"`, `"architect"`, `"leonids"`, and `"hpstr"`. Figure 9.1 shows the appearance of the `leonids` theme. See `https://github.com/yixuan/prettydoc` for screenshots of more themes.

- The `highlight` option takes `null`, `"github"`, or `"vignette"`.

Below is an example of the YAML metadata of an R Markdown document that uses the `prettydoc::html_pretty` output format:

```
---
title: "Your Document Title"
author: "Document Author"
date: "2018-04-16"
output:
  prettydoc::html_pretty:
    theme: leonids
    highlight: github
---
```

9.1.2 Package vignettes

The `prettydoc::html_pretty` can be particularly useful for R package vignettes. We have mentioned the `html_vignette` format in Section 3.8 that also aims at smaller file sizes, but that format is not as stylish. To apply the `prettydoc::html_pretty` format to a package vignette, you may use the YAML metadata below:

```
---
title: "Vignette Title"
author: "Vignette Author"
output: prettydoc::html_pretty
vignette: >
  %\VignetteIndexEntry{Vignette Title}
  %\VignetteEngine{knitr::rmarkdown}
```

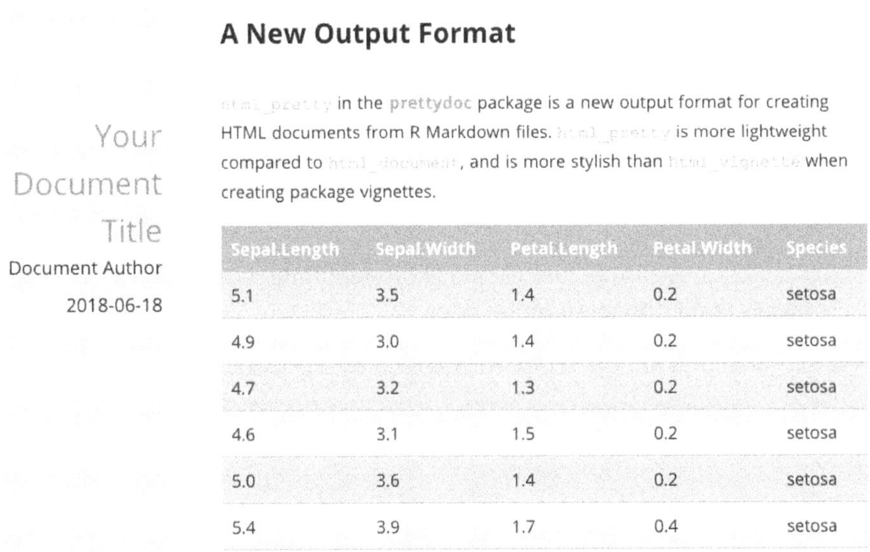

A New Output Format

Your
Document
Title

Document Author

2018-06-18

html_pretty in the prettydoc package is a new output format for creating HTML documents from R Markdown files. html_pretty is more lightweight compared to html_document, and is more stylish than html_vignette when creating package vignettes.

Sepal.Length	Sepal.Width	Petal.Length	Petal.Width	Species
5.1	3.5	1.4	0.2	setosa
4.9	3.0	1.4	0.2	setosa
4.7	3.2	1.3	0.2	setosa
4.6	3.1	1.5	0.2	setosa
5.0	3.6	1.4	0.2	setosa
5.4	3.9	1.7	0.4	setosa

FIGURE 9.1: The leonids theme of the prettydoc package.

```
%\VignetteEncoding{UTF-8}
---
```

Do not forget to change the vignette title, author, and the index entry. You should also add prettydoc to the Suggests field of your package DESCRIPTION file, and the two package names knitr, rmarkdown to the VignetteBuilder field.

9.2 The rmdformats package

The **rmdformats** package (Barnier, 2017) provides several HTML output formats of unique and attractive styles, including:

- material: A format based on the Material design theme for Bootstrap 3[1]. With this format, every first-level section will become a separate page. See

---

[1]https://github.com/FezVrasta/bootstrap-material-design

Figure 9.2 for what this format looks like ("Introduction" and "Including Plots" are two first-level sections).

- `readthedown`: It features a sidebar layout. The table of contents is displayed in the sidebar on the left. As you scroll on the page, the current section header will be automatically highlighted (and expanded if necessary) in the sidebar.

- `html_clean`: A simple and clean HTML template, with a dynamic table of contents at the top-right of the page.

- `html_docco`: A simple template inspired by the Docco project[2].

Do not forget the `rmdformats::` prefix when you use these formats, e.g.,

```
---
output: rmdformats::material
---
```

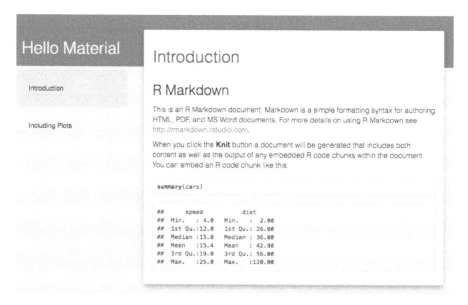

FIGURE 9.2: The Material Design theme in the rmdformats package.

These output formats have some additional features such as responsiveness and code folding. Please refer to the GitHub repository of the **rmdformats** package for more information: `https://github.com/juba/rmdformats`.

---

[2]`https://github.com/jashkenas/docco`

9.3 Shower presentations

Shower (`https://github.com/shower/shower`) is a popular and customizable HTML5 presentation framework. See Figure 9.3 for what it looks like.

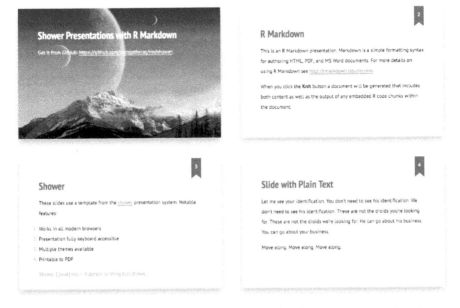

FIGURE 9.3: A few sample slides created via the Shower presentation framework.

The R package **rmdshower** (`https://github.com/mangothecat/rmdshower`) is built on top of Shower. You may install it from GitHub:

```
devtools::install_github("mangothecat/rmdshower")
```

You can create a Shower presentation with the output format `rmdshower::shower_presentation`, e.g.,

```
---
title: "Hello Shower"
author: "John Doe"
output: rmdshower::shower_presentation
---
```

See the help page `?rmdshower::shower_presentation` for all possible options of this format.

10

Websites

Most R Markdown applications are single documents. That is, you have a single R Markdown source document, and it generates a single output file. However, it is also possible to work with multiple Rmd documents in a project, and organize them in a meaningful way (e.g., pages can reference each other).

Currently there are two major ways to build multiple Rmd documents: **blogdown** (Xie et al., 2017; Xie, 2018a) for building websites, and **bookdown** (Xie, 2016, 2018b) for authoring books. In this chapter, we briefly introduce the **blogdown** package. For the full reference, please read the official **blogdown** book by Xie et al. (2017). In fact, the **rmarkdown** package also has a built-in site generator, which was not covered in detail by the **blogdown** book, so we will introduce it in Section 10.5.

With **blogdown**, you can write a blog post or a general page in an Rmd document, or a plain Markdown document. These source documents will be built into a static website, which is essentially a folder containing static HTML files and associated assets (such as images and CSS files). You can publish this folder to any web server as a website. Because it is only a single folder, it can be easy to maintain. For example, you do not need to worry about databases as you do if you use dynamic systems like WordPress.

Because the website is generated from R Markdown, the content is more likely to be reproducible, and also easier to maintain (no cut-and-paste of results). Using Markdown means your content could be more portable in the sense that you may convert your pages to PDF or other formats in the future, and you are not tied to the default HTML format. For example, you may be able to convert a blog post to a journal paper, or several posts to a book. One more benefit of using **blogdown** is that the Markdown syntax is based on **bookdown**'s extended syntax, which means it is highly suitable for technical writing. For example, you may write math equations, insert figures or tables with captions, cross-reference them with figure or table numbers, add citations, and present theorems or proofs.

10.1   Get started

You can install **blogdown** from CRAN. If you want to test the development version, you may also install it from GitHub:

```
# from CRAN
install.packages("blogdown")

# or the development version from GitHub
devtools::install_github("rstudio/blogdown")
```

The easiest way to get started with a **blogdown**-based website is to create a website project from RStudio: File -> New Project. If you do not use RStudio, you may call the function blogdown::new_site().

The first time when you create a new website, **blogdown** will do a series of things behind the scenes: it downloads Hugo (the default static site generator), creates a website skeleton, installs a theme, adds some example posts, builds the site, and serves it so that you can see the website in your browser (or RStudio Viewer if you are in RStudio). It will not go through all these steps again the next time when you work on this website. All you need in the future is blogdown::serve_site(), or equivalently, the RStudio addin "Serve Site".

Every time you open a website project, you only need to serve the site once, and **blogdown** will keep running in the background, listening to changes in your source files, and rebuilding the website automatically. All you have to do is create new posts, or edit existing posts, and save them. You will see the automatic live preview as you save the changes (unless you have errors in a source document).

There are a few RStudio addins to help you author your posts: you can use the "New Post" addin to create a new post, the "Update Metadata" addin to update the YAML metadata of a post, and the "Insert Image" addin to insert an image in a post.

10.2   The directory structure

The default site generator in **blogdown** is Hugo (`https://gohugo.io`). A basic Hugo website usually contains the following files and directories:

- `config.toml`
- `content/`
- `static/`
- `themes/`
- `public/`

The configuration file `config.toml` can be used to specify options for the Hugo website, e.g.,

```
baseURL = "/"
languageCode = "en-us"
title = "A Hugo website"
theme = "hugo-lithium-theme"
```

Some options are provided by Hugo itself, such as `title` and `baseURL`; you may refer to `https://gohugo.io/getting-started/configuration/` for all built-in options. Some options are provided by the Hugo theme, and you need to read the documentation of the specific theme to know the additional options.

All source Markdown or R Markdown files should be placed under the `content/` directory. The directory structure under `content/` can be arbitrary.

The `static/` directory contains static assets such as images and CSS files. Everything under `static/` will be copied to the `public/` directory when Hugo generates the website. For example, `static/images/foo.png` will be copied to `public/images/foo.png`, and if you want to include this image in your post, you may use `![title](/images/foo.png)` in Markdown (the leading `/` typically indicates the root of `public/`).

You can download multiple themes to the `themes` directory. To activate a theme, specify its folder name in the `theme` option in `config.toml`. You can find a lot of Hugo themes from `https://themes.gohugo.io`. Remember, the best theme is always the next one, i.e., one that you have not used before. I recommend that you start with a simple theme (such as the default hugo-lithium theme[1] in **blogdown**, hugo-xmin[2], or hugo-tanka[3]), and write a substantial number of posts before seriously investing time in choosing or tweaking a theme.

After you serve a site using **blogdown**, your site will be continuously built to the `public/` directory by default. You can upload this folder to any web server to deploy the website. However, if you know GIT, there is an even easier way to publish your website, to be introduced in the next section.

Hugo is very powerful and customizable. If you want to learn more technical details about it, you may read Chapter 2 of the **blogdown** book.

10.3 Deployment

There are multiple ways to deploy a website, such as using your own web server, GitHub Pages, or Amazon S3. We only mention one in this chapter: Netlify (`https://www.netlify.com`). It provides both free and paid plans. For personal users, the free plan may be enough, because many useful features have been included in the free plan, e.g., the Hugo support, CDN (content delivery network) for high availability and performance of your website, HTTPS, binding your custom domain, and 301/302 redirects.

Netlify currently supports GitHub, GitLab, and Bitbucket. You may log in using one of these accounts at `https://app.netlify.com`, and create a new website from your GIT repository that contains the source of your website. Note that you do not need to commit or push the `public/` directory in GIT (in fact, I recommend that you ignore this directory in `.gitignore`).

When creating a new site on Netlify, you can specify the build command to be `hugo`, the publish directory to be `public` (unless you changed the setting `publishDir` in `config.toml`), and also add an environment variable `HUGO_VERSION` with a value of a suitable Hugo version (e.g., `0.39`). To find the Hugo version on your local computer, call the function `blogdown::hugo_version()`. You may want to use the same Hugo version on Netlify.

---

[1]`https://github.com/yihui/hugo-lithium`
[2]`https://github.com/yihui/hugo-xmin`
[3]`https://github.com/road2stat/hugo-tanka`

Netlify will assign a random subdomain of the form xxx-xxx-1234.netlify.com to you. You may change it to a meaningful domain name, or request a free *.rbind.io domain name from https://github.com/rbind/support/issues if you like it.

If possible, I strongly recommend that you enable HTTPS for your websites (why?[4]). HTTPS is free on Netlify, so you really do not have a reason not to enable it.

Once your GIT repository is connected with Netlify, you only need to push source files to the repository in the future, and Netlify will automatically rebuild your website. This is called "continuous deployment".

10.4 Other site generators

Currently **blogdown** has limited support for two other popular site generators: Jekyll[5] and Hexo[6]. You can find detailed instructions on how to configure **blogdown** for these site generators in Chapter 5 of the **blogdown** book. Note that neither Pandoc's Markdown nor HTML widgets are supported if you use Jekyll or Hexo with **blogdown**.

10.5 rmarkdown's site generator

Before **blogdown** was invented, the **rmarkdown** package had provided a simple site generator that did not rely on a third-party site generator like Hugo. If you feel Hugo is too complex for you, and you only want to build a few Rmd documents into a website, this built-in site generator may be a good choice. A main restriction of this site generator is that it assumes all Rmd documents are under a flat directory (i.e., no pages under subdirectories). It also has fewer features compared to Hugo (e.g., no RSS feeds).

---

[4]https://https.cio.gov/everything/
[5]https://jekyllrb.com
[6]https://hexo.io

You can render collections of R Markdown documents as a website using the `rmarkdown::render_site()` function. We will call such websites "R Markdown websites" in this section. The RStudio IDE (version 1.0 or higher) also includes integrated support for developing R Markdown websites.

10.5.1 A simple example

To start with, let's walk through a very simple example, a website that includes two pages (`Home` and `About`) and a navigation bar to switch between them.

First, we need a configuration file `_site.yml`:

```
name: "my-website"
navbar:
  title: "My Website"
  left:
    - text: "Home"
      href: index.html
    - text: "About"
      href: about.html
```

Then two Rmd files, `index.Rmd`:

```
---
title: "My Website"
---

Hello, Website!
```

and `about.Rmd`:

```
---
title: "About This Website"
---

More about this website.
```

Note that the minimum requirement for any R Markdown website is that it have an `index.Rmd` file as well as a `_site.yml` file. If you execute the

`rmarkdown::render_site()` function from within the directory containing
the website, the following will occur:

1. All of the `*.Rmd` and `*.md` files in the root website directory will
   be rendered into HTML. Note, however, that Markdown files beginning with _ are not rendered (this is a convention to designate
   files that are to be included by top level Rmd documents as child
   documents).

2. The generated HTML files and any supporting files (e.g., CSS and
   JavaScript) are copied into an output directory (`_site` by default).

The HTML files within the `_site` directory are now ready to deploy as a
standalone static website.

The full source code for the simple example above can be found in the `hello-
website` folder in the repository `https://github.com/rstudio/rmarkdown-
website-examples`.

10.5.2 Site authoring

10.5.2.1 RStudio

RStudio includes a variety of features intended to make developing R Markdown websites more productive.

All of the RStudio features for website authoring described below require the
use of an RStudio Project tied to your website's directory. See the documentation on RStudio Projects[7] for additional information on how to create and
use projects.

As you work on the individual pages of your website, you can render them
using the `Knit` button just as you do with conventional standalone R Markdown documents (see Figure 10.1).

Knitting an individual page will only render and preview that page, not the
other pages in the website.

To render all of the pages in the website, you use the `Build` pane, which calls
`rmarkdown::render_site()` to build and then preview the entire site (see
Figure 10.2).

---

[7]`https://support.rstudio.com/hc/en-us/articles/200526207-Using-Projects`

FIGURE 10.1: Knit a single page of a website.

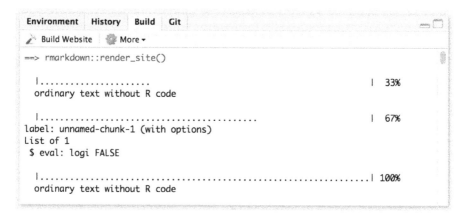

FIGURE 10.2: Build an entire website in RStudio.

RStudio supports "live preview" of changes that you make to supporting files within your website (e.g., CSS, JavaScript, Rmd partials, R scripts, and YAML config files).

Changes to CSS and JavaScript files always result in a refresh of the currently active page preview. Changes to other files (e.g., shared scripts and configuration files) trigger a rebuild of the active page (this behavior can be disabled via the options dialog available from the Build pane).

Note that only the active page is rebuilt, so once you are happy with the results of rendering you should make sure to rebuild the entire site from the Build pane to ensure that all pages inherit your changes.

When working iteratively on a page, you might find it more convenient to preview it side-by-side with the editor rather than in an external window. You can configure RStudio to do this using the options menu on the editor toolbar (see Figure 10.3).

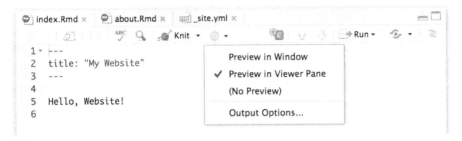

FIGURE 10.3: Preview a page side-by-side with the editor in RStudio.

10.5.2.2 Command line

If you are not working within RStudio and/or want to build sites from the command line, you can call the render_site() function directly from within the website directory. Pass no arguments to render the entire site or a single file in order to render just that file:

```
# render the entire site
rmarkdown::render_site()
```

```
# render a single file only
rmarkdown::render_site("about.Rmd")
```

To clean up all of the files generated via render_site(), you can call the clean_site() function, which will remove all files generated by rendering your site's Markdown documents, including **knitr**'s *_cache directories. You can specify the preview = FALSE option to just list the files to be removed rather than actually removing them:

```
# list which files will be removed
rmarkdown::clean_site(preview = TRUE)
```

```
# actually remove the files
rmarkdown::clean_site()
```

10.5.2.3   knitr caching

If your website is time consuming to render, you may want to enable **knitr**'s caching during the development of the site, so that you can more rapidly preview. To enable caching for an individual chunk, just add the cache = TRUE chunk option:

````
```{r, cache = TRUE}
data <- longComputation()
```
````

To enable caching for an entire document, add cache = TRUE to the global chunk option defaults:

````
```{r setup, include=FALSE}
knitr::opts_chunk$set(cache = TRUE)
```
````

Note that when caching is enabled for an Rmd document, its *_files directory will be copied rather than moved to the _site directory (since the cache requires references to generated figures in the *_files directory).

10.5.3   Common elements

10.5.3.1   Content

Typically when creating a website, there are various common elements you want to include on all pages (e.g., output options, CSS styles, header and footer elements, etc.). Here are additions in three files to the example above to make use of common elements:

- _site.yml:

```yaml
name: "my-website"
navbar:
  title: "My Website"
  left:
    - text: "Home"
      href: index.html
```

```
    - text: "About"
      href: about.html
  output:
    html_document:
      theme: cosmo
      highlight: textmate
      include:
        after_body: footer.html
      css: styles.css
```

- `footer.html`:

```
<p>Copyright &copy; 2016 Skynet, Inc. All rights reserved.</p>
```

- `styles.css`

```
blockquote {
  font-style: italic
}
```

Note that we have included an `output` element within our `_site.yml` file. This defines shared output options for all R Markdown documents within a site. Note that individual documents can also include their own `output` options, which will be merged with the common options at render time.

As part of our common output options, we have specified an HTML footer (via the `include: after-body:` option) and a CSS stylesheet. You can also include HTML before the body or in the document's `<head>` tag (see Section 3.1.10.2).

In addition to whatever common options you define, there are two output options that are automatically set when rendering a site:

1. The `self_contained` option is set `FALSE`; and

2. The `lib_dir` option is set to `site_libs`.

These options are set so that dependent files (e.g., jQuery, Bootstrap, and HTML widget libraries) are shared across all documents within the site rather than redundantly embedded within each document.

10.5.3.2   R scripts

If you have R code that you would like to share across multiple R Markdown documents within your site, you can create an R script (e.g., utils.R) and source it within your Rmd files. For example:

```r
source("utils.R")
```

10.5.3.3   Rmd partials

You may have common fragments of R Markdown that you want to share across pages within your site. To share Rmd fragments, you should name them with a leading underscore (_), and then include them within their parent Rmd document using the child chunk option. For example:

- about.Rmd:

  ```
  ---
  title: "About This Website"
  ---

  More about this website.
  ```

  ```{r, child="_session-info.Rmd"}
  ```

- _session-info.Rmd:

  ```
  Session information:
  ```

  ```{r}
  sessionInfo()
  ```

The leading underscore is an indicator to the site generation engine that the Rmd is a partial document to be included in other documents, so it is not compiled as a standalone document during site rendering.

The full source code for the above example can be found in the `common-element` folder in the repository `https://github.com/rstudio/rmarkdown-website-examples`.

10.5.4 Site navigation

The `navbar` element of `_site.yml` can be used to define a common navigation bar for your website. You can include internal and external links on the navigation bar as well as drop-down menus for sites with a large number of pages.

Here is a navigation bar definition in `_site.yml` that makes use of a variety of features:

```yaml
name: "my-website"
navbar:
  title: "My Website"
  type: inverse
  left:
    - text: "Home"
      icon: fa-home
      href: index.html
    - text: "About"
      icon: fa-info
      href: about.html
    - text: "More"
      icon: fa-gear
      menu:
        - text: "Heading 1"
        - text: "Page A"
          href: page-a.html
        - text: "Page B"
          href: page-b.html
        - text: "---------"
        - text: "Heading 2"
        - text: "Page C"
          href: page-c.html
        - text: "Page D"
          href: page-d.html
```

```
right:
  - icon: fa-question fa-lg
    href: https://example.com
```

This example demonstrates a number of capabilities of navigation bars:

1. You can use the `type` field to choose between the `default` and `inverse` navigation bar styles (each theme includes distinct colors for "default" and "inverse" navigation bars).

2. You can align navigational items either to the `left` or to the `right`.

3. You can include menus on the navigation bar, and those menus can have separators (`text: "--------------"`) and internal headings (`text` without a corresponding `href`).

4. You can include both internal and external links on the navigation bar.

5. You can use icons on the navigation bar. Icons from three different icon sets are available.

   - Font Awesome[8]
   - Ionicons[9]
   - Bootstrap Glyphicons[10]

   When referring to an icon, you should use its full name including the icon set prefix (e.g., `fa-github`, `ion-social-twitter`, and `glyphicon-time`).

10.5.4.1 HTML navigation bar

If you want to have even more control over the appearance and behavior of the navigation bar, you can define it in HTML rather than YAML. If you include a file named `_navbar.html` in your website directory, it will be used as the navigation bar. Here is an example of navigation bar defined in HTML: `https://github.com/rstudio/rmarkdown-website/blob/master/_navbar.html`.

---

[8] `https://fontawesome.com/icons`
[9] `http://ionicons.com/`
[10] `https://getbootstrap.com/components/`

Full documentation on syntax of Bootstrap navigation bars can be found here:
`http://getbootstrap.com/components/`.

10.5.5 HTML generation

R Markdown includes many facilities for generation of HTML content from R objects, including:

- The conversion of standard R output types (e.g., textual output and plots) within code chunks done automatically by **knitr**.

- A variety of ways to generate HTML tables, including the `knitr::kable()` function and other packages such as **kableExtra** and **pander**.

- A large number of available HTML widgets that provide rich JavaScript data visualizations.

As a result, for many R Markdown websites you will not need to worry about generating HTML output at all (since it is created automatically).

10.5.5.1 The htmltools package

If the facilities described above do not meet your requirements, you can also generate custom HTML from your R code using the **htmltools** package (RStudio and Inc., 2017). The **htmltools** package enables you to write HTML using a convenient R based syntax (this is the same core HTML generation facility used by the **shiny** package).

Here is an example of an R function that creates a Bootstrap thumbnail div:

```r
library(htmltools)
thumbnail <- function(title, img, href, caption = TRUE) {
  div(class = "col-sm-4",
      a(class = "thumbnail", title = title, href = href,
        img(src = img),
        div(class = if (caption) "caption",
            if (caption) title)
      )
  )
}
```

You can write functions that build HTML like the one above, then call them from other R code that combines them with your data to produce dynamic HTML. An R code chunk that makes use of this function might look like this:

```{r, echo=FALSE}
thumbnail("Apple", "images/apple.png",
          "https://en.wikipedia.org/wiki/Apple")
thumbnail("Grape", "images/grape.png",
          "https://en.wikipedia.org/wiki/Grape")
thumbnail("Peach", "images/peach.png",
          "https://en.wikipedia.org/wiki/Peach")
```

10.5.6 Site configuration

The _site.yml file has a number of options that affect site output, including where it is written and what files are included and excluded from the site. Here is an example that makes use of a few of these options:

```
name: "my-website"
output_dir: "_site"
include: ["import.R"]
exclude: ["docs.txt", "*.csv"]
```

The name field provides a suggested URL path for your website when it is published (by default this is just the name of the directory containing the site).

The output_dir field indicates which directory to copy site content into ("_site" is the default if none is specified). It can be "." to keep all content within the root website directory alongside the source code.

10.5.6.1 Included files

The include and exclude fields enable you to override the default behavior vis-a-vis what files are copied into the output directory. By default, all files within the website directory are copied into the output directory save for the following:

1. Files beginning with . (hidden files).

2. Files beginning with _.

3. Files known to contain R source code (e.g., `*.R`, `*.s`, `*.Rmd`), R data (e.g., `*.RData`, `*..rds`), or configuration data (e.g., `*..Rproj`, `rsconnect`).

The `include` and `exclude` fields of `_site.yml` can be used to override this default behavior (wildcards can be used to specify groups of files to be included or excluded).

Note that `include` and `exclude` are *not* used to determine which Rmd files are rendered: all of them in the root directory save for those named with the _ prefix will be rendered.

10.5.7 Publishing websites

R Markdown websites are static HTML sites that can be deployed to any standard web server. All site content (generated documents and supporting files) are copied into the `_site` directory, so deployment is simply a matter of moving that directory to the appropriate directory of a web server.

10.5.8 Additional examples

Here are some additional examples of websites created with R Markdown:

- The **rmarkdown** documentation: `https://rmarkdown.rstudio.com`. This website was created using R Markdown. There are a large number of pages (over 40) that are organized using sub-menus on the navigation bar. Disqus comments are included on each page via an `after_body` option. The source code is at `https://github.com/rstudio/rmarkdown/tree/gh-pages`.

- The **flexdashboard** documentation: `https://rmarkdown.rstudio.com/flexdashboard/`. It illustrates using an R script to dynamically generate HTML thumbnails of **flexdashboard** examples from YAML. The source code is at `https://github.com/rstudio/rmarkdown/tree/gh-pages/flexdashboard`.

10.5.9 Custom site generators

So far we have described the behavior of the default site generation function, `rmarkdown::default_site()`. It is also possible to define a custom site generator that has alternate behaviors.

10.5.9.1 Site generator function

A site generator is an R function that is bound to by including it in the `site:` field of the `index.Rmd` or `index.md` file. For example:

```
---
title: "My Book"
output: bookdown::gitbook
site: bookdown::bookdown_site
---
```

A site generation function should return a list with the following elements:

- `name`: The name for the website (e.g., the parent directory name).
- `output_dir`: The directory where the website output is written to. This path should be relative to the site directory (e.g., "." or "_site").
- `render`: An R function that can be called to generate the site. The function should accept the `input_file`, `output_format`, `envir`, `quiet`, and `encoding` arguments.
- `clean`: An R function that returns relative paths to the files generated by `render_site()`. These files are the ones to be removed by the `clean_site()` function.

Note that the `input_file` argument will be `NULL` when the entire site is being generated. It will be set to a specific filename if a front-end tool is attempting to preview it (e.g., RStudio IDE via the `Knit` button).

When `quiet = FALSE`, the `render` function should also print a line of output using the `message()` function indicating which output file should be previewed. For example:

```
if (!quiet)
  message("\nOutput created: ", output)
```

Emitting this line enables front-ends like RStudio to determine which file they should open to preview the website.

10.5.9.2 Examples

See the source code of the `rmarkdown::default_site` function for an example of a site generation function. The **bookdown** package also implements a custom site generator via its `bookdown::bookdown_site` function.

11

HTML Documentation for R Packages

R has a built-in HTML help system that can be accessed via `help.start()`. From this system, you can see the HTML help pages of functions and objects in all packages, as well as other information about packages such as the `DESCRIPTION` file and package vignettes. However, this system is usually dynamically launched (via a local web server), and it is not straightforward to turn it into a static website that can be viewed without starting R.

The **pkgdown** package (Wickham and Hesselberth, 2018) makes it easy to build a documentation website for an R package, which can help you organize different pieces of the package documentation (e.g., README, help pages, vignettes, and news) with a more visually pleasant style. The navigation can also be easier for users than R's built-in help system. This website can be published to any web server (e.g., GitHub Pages or Netlify). An example is **pkgdown**'s own website: `http://pkgdown.r-lib.org` (see Figure 11.1).

11.1 Get started

You can install **pkgdown** from CRAN, or its development version from GitHub, and find more information from its GitHub repository (`https://github.com/r-lib/pkgdown`).

```r
install.packages("pkgdown")

# Or the development version
devtools::install_github("r-lib/pkgdown")
```

After it is installed, you can call the function `pkgdown::build_site()` in the

FIGURE 11.1: A screenshot of the pkgdown website.

root directory of your source package. It will build a website to the docs/ directory, which can be turned into an online website via GitHub Pages[1] or Netlify.

11.2   Components

A **pkgdown** website consists of these components: the home page, function reference, articles, news, and the navigation bar. You may configure these components via a file _pkgdown.yml.

---

[1]https://help.github.com/articles/configuring-a-publishing-source-for-github-pages/

11.2.1 Home page

The home page is generated from the first existing file of the following files in your source package:

- `index.Rmd`
- `README.Rmd`
- `index.md`
- `README.md`

Other meta information about the package, such as the package license and author names, will be displayed automatically as a sidebar on the home page.

11.2.2 Function reference

The reference pages look like R's own help pages. In fact, these pages are generated from the `*.Rd` files under `man/`. Compared to R's own help pages, **pkgdown** offers a few more benefits: the examples on a help page (if they exist) will be evaluated so that you can see the output, and function names are automatically linked so you can click on a name to navigate to the help page of another function. What is more, **pkgdown** allows you to organize the list of all functions into groups (e.g., by topic), which can make it easier for users to find the right function in a list. By default, all functions are listed alphabetically just like R's help system. To group functions on the list page, you need to provide a `reference` key in `_pkgdown.yml`, e.g.,

```
reference:
  - title: "One Topic"
    desc: "These functions are awesome..."
    contents:
      - awesome_a
      - awesome_b
      - cool_c
  - title: "Another Topic"
    desc: "These functions are boring..."
    contents:
      - starts_with("boring_")
      - ugh_oh
```

As you can see from the above example, you may list the names of functions in the `contents` field, or provide a pattern to let **pkgdown** match the names. There are three ways to match function names: `starts_with()` to match names that start with a string, `ends_width()` for an ending pattern, and `matches()` for an arbitrary regular expression.

11.2.3 Articles

Package vignettes in the R Markdown format under the `vignettes/` directories will be built as "articles" for a **pkgdown**-based website. Note that Rmd files under subdirectories will also be built. The list of articles will be displayed as a drop-down menu in the navigation bar.

If you have a vignette that has the same base name as the package name (e.g., a vignette `foo.Rmd` in a package **foo**), it will be displayed as the "Get started" menu item in the navigation bar.

11.2.4 News

If the source package has a news file `NEWS.md`, it will be parsed and rendered to HTML pages that can be accessed via the "Changelog" menu in the navigation bar.

11.2.5 Navigation bar

The navigation bar in **pkgdown** is based on the **rmarkdown** site generator. You can learn how to customize it from Section 10.5.4, if you are not satisfied by the default navigation bar. Please note that you need to specify the `navbar` field in `_pkgdown.yml` instead of `_site.yml`.

12

Books

We have introduced the basics of R Markdown in Chapter 3, which highlighted how HTML, PDF, and Word documents can be produced from an R workflow. However, larger projects can become difficult to manage in a single R Markdown file. The **bookdown** package (Xie, 2016, 2018b) addresses this limitation, and offers several key improvements:

- Books and reports can be built from multiple R Markdown files.

- Additional formatting features are added, such as cross-referencing, and numbering of figures, equations, and tables.

- Documents can easily be exported in a range of formats suitable for publishing, including PDF, e-books and HTML websites.

This book itself was created using **bookdown**, and acts as an example of what can be achieved. Despite the name containing the word "book", **bookdown** is not only for books, and it can be used for long reports, dissertations, or even single R Markdown documents (see Section 12.4.4). It also works with other computing languages such as Python and C++ (see Section 2.7). If you want, you can even write documents irrelevant to computing, such as a novel.

In this chapter, we cover the basics of **bookdown**, and explain how to start a **bookdown** project. Much of the the content is based on the work *"bookdown: Authoring Books and Technical Documents with R Markdown"* (https://bookdown.org/yihui/bookdown/) of Xie (2016), which provides more detailed explanations of the concepts highlighted.

12.1   Get started

You can install either the CRAN version or the development version on
GitHub (`https://github.com/rstudio/bookdown`):

```
# install from CRAN
install.packages("bookdown")

# or GitHub
devtools::install_github("rstudio/bookdown")
```

If you use RStudio, you can start a new bookdown project from the menu
`File -> New Project -> New Directory -> Book Project using book-`
`down`.[1] Open the R Markdown file `index.Rmd`, and click the button `Build`
`Book` on the `Build` tab of RStudio. This will compile the book and display
the HTML version within the RStudio Viewer, which looks like Figure 12.1.

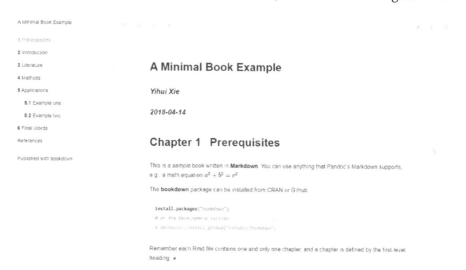

FIGURE 12.1: The HTML output of the bookdown template.

You may add or change the R Markdown files, and hit the `Knit` button again

---

[1]Alternatively, the command `bookdown:::bookdown_skeleton(getwd())` will create a
skeleton project in your current working directory.

to preview the book. If you prefer not to use RStudio, you may also compile the book through the command line using `bookdown::render_book()`.

12.2   Project structure

Below shows the basic structure of a default **bookdown** project:

```
directory/
├──   index.Rmd
├── 01-intro.Rmd
├── 02-literature.Rmd
├── 03-method.Rmd
├── 04-application.Rmd
├── 05-summary.Rmd
├── 06-references.Rmd
├── _bookdown.yml
├── _output.yml
├──   book.bib
├──   preamble.tex
├──   README.md
└──   style.css
```

As a summary of these files:

- `index.Rmd`: This is the only Rmd document to contain a YAML frontmatter as described within Chapter 2, and is the first book chapter.

- Rmd files: A typical **bookdown** book contains multiple chapters, and one chapter lives in one Rmd file.

- `_bookdown.yml`: A configuration file for **bookdown**.

- `_output.yml`: It specifies the formatting of the HTML, LaTeX/PDF, and e-books.

- `preamble.tex` and `style.css`: They can be used to adjust the appearance and styles of the book output document(s). Knowledge of LaTeX and/or CSS is required.

These files are explained in greater detail in the following subsections.

12.2.1   Index file

By default, all Rmd files are merged to render the book. The `index.Rmd` file is the first file when merging all Rmd files. You should and should only specify the YAML metadata in this file, e.g.,

```
---
title: "A Minimal Book Example"
author: "Yihui Xie"
date: "`r Sys.Date()`"
site: bookdown::bookdown_site
documentclass: book
bibliography: [book.bib, packages.bib]
biblio-style: apalike
link-citations: yes
description: "This is a minimal example of using
  the bookdown package to write a book."
---
```

12.2.2   Rmd files

The rest of Rmd files must start immediately with the chapter title using the first-level heading, e.g., `# Chapter Title`.

- 01-intro.Rmd

  ```
  # Introduction

  This chapter is an overview of the methods that
  we propose to solve an **important problem**.
  ```

- 02-literature.Rmd

  ```
  # Literature

  Here is a review of existing methods.
  ```

By default, **bookdown** merges all Rmd files by the order of filenames, e.g., `01-intro.Rmd` will appear before `02-literature.Rmd`. Filenames that start with an underscore _ are skipped.

12.2.3   `_bookdown.yml`

The `_bookdown.yml` file allows you to specify optional settings to build the book. For example, you may want to override the order in which files are merged by including the field `rmd_files`:

```
rmd_files: ["index.Rmd", "02-literature.Rmd", "01-intro.Rmd"]
```

12.2.4   `_output.yml`

The `_output.yml` file is used to specify the book output formats (see Section 12.4). Here is a brief example:

```
bookdown::gitbook:
  lib_dir: assets
  split_by: section
  config:
    toolbar:
      position: static
bookdown::pdf_book:
  keep_tex: yes
bookdown::html_book:
  css: toc.css
```

12.3   Markdown extensions

The **bookdown** package expands upon the Markdown syntax outlined in Section 2.5, and provides additional powerful features that assist longer documents and academic writing.

12.3.1 Number and reference equations

Section 2.5.3 highlighted how equations can be created using LaTeX syntax within Markdown. To number equations, put them in the `equation` environments, and assign labels to them using the syntax `(\#eq:label)`. Equation labels must start with the prefix `eq:` in **bookdown**. For example:

```
\begin{equation}
  E=mc^2
  (\#eq:emc)
\end{equation}
```

It renders the equation below (12.1):

$$E = mc^2 \tag{12.1}$$

12.3.2 Theorems and proofs

Theorems and proofs provide environments that are commonly used within articles and books in mathematics. To write a theorem, you can use the syntax below:

```
```{theorem}
Here is my theorem.
```
```

For example:

Theorem 12.1 (Pythagorean theorem). *For a right triangle, if c denotes the length of the hypotenuse and a and b denote the lengths of the other two sides, we have*

$$a^2 + b^2 = c^2$$

Theorems can be numbered and cross-referenced, as you can see from Theorem 12.1. The `proof` environment behaves similarly to theorem environments but is unnumbered.

Variants of the `theorem` environments include: `lemma`, `corollary`, `proposition`, `conjecture`, `definition`, `example`, and `exercise`. Variants of the `proof` environments include `remark` and `solution`. The syntax for these environments is similar to the `theorem` environment, e.g., ```` ``` ````{lemma}.

12.3.3 Special headers

There are two special types of first-level headers than can be used in **bookdown**:

- A part can be created using `# (PART) Part Title {-}` before the chapters that belong to this part.

- Appendices `# (APPENDIX) Appendix {-}`: All chapters after this header will be treated as the appendix. The numbering style of these chapters will be A, B, C, etc., and sections will be numbered as `A.1`, `A.2`, and so on.

12.3.4 Text references

A text reference is a paragraph with a label. The syntax is `(ref:label) text`, where `label` is a unique identifier, and `text` is a Markdown paragraph. For example:

```
(ref:foo) Define a text reference **here**.
```

Then you can use `(ref:foo)` to refer to the full text. Text references can be used anywhere in the document, and are particularly useful when assigning a long caption to a figure or including Markdown formatting in a caption. For example:

```
Some text.

(ref:cool-plot) A boxplot of the data `iris` in **base** R.

```{r cool-plot, fig.cap='(ref:cool-plot)'}
boxplot(Sepal.Length ~ Species, data = iris)
```
```

12.3.5 Cross referencing

The **bookdown** package extends cross-referencing in R Markdown documents and allows section headers, tables, figures, equations, and theorems to be cross-referenced automatically. This only works for numbered environments, and therefore requires figures and tables to be assigned a label. Cross-references are made in the format \@ref(type:label), where label is the chunk label and type is the environment being referenced. As examples:

- Headers:

  ```
  # Introduction {#intro}

  This is Chapter \@ref(intro)
  ```

- Figures:

  ```
  See Figure \@ref(fig:cars-plot)

  ```{r cars-plot, fig.cap="A plot caption"}
  plot(cars)  # a scatterplot
  ```
  ```

- Tables:

  ```
  See Table \@ref(tab:mtcars)

  ```{r mtcars}
  knitr::kable(mtcars[1:5, 1:5], caption = "A caption")
  ```
  ```

- Theorems:

  ```
  See Theorem \@ref(thm:boring)

  ```{theorem, boring}
  Here is my theorem.
  ```
  ```

- Equations:

```
See equation \@ref(eq:linear)

\begin{equation}
a + bx = c   (\#eq:linear)
\end{equation}
```

Note that only alphanumeric characters (a-z, A-Z, 0-9), -, /, and : are allowed in these labels.

12.4 Output Formats

The **bookdown** package includes the following output formats:

- HTML:
  - gitbook
  - html_book
  - tufte_html_book
- PDF:
  - pdf_book
- e-book:
  - epub_book
- Single documents:
  - html_document2
  - tufte_html2
  - pdf_document2
  - tufte_handout2
  - tufte_book2
  - word_document2

12.4.1   HTML

Although multiple formats are available for HTML books in **bookdown**, we will focus on the Gitbook style, which appears to be the most popular format. It provides a clean style, with a table of contents on the left. The design is fully responsive to make the content suitable for both mobile and desktop devices.

The output format `bookdown::gitbook` is built upon `rmark-down::html_document`, which was explained in Section 3.1. The main difference between rendering in R Markdown and **bookdown** is that a book will generate multiple HTML pages by default. To change the way the HTML pages are split, the `split_by` argument can be specified. This defaults to `split_by: chapter`, but readers may prefer to use `split_by: section` if there are many sections within chapters, in which case a chapter page may be too long.

12.4.2   LaTeX/PDF

There are limited differences between the output of `pdf_book()` in **bookdown** compared to `pdf_document()` in **rmarkdown**. The primary purpose of the new format is to resolve the labels and cross-references written in the syntax described in Section 12.3.5.

Pandoc supports LaTeX commands in Markdown. Therefore if the only output format that you want for a book is LaTeX/PDF, you may use the syntax specific to LaTeX, such as `\newpage` to force a page break. A major disadvantage of this approach is that LaTeX syntax is not portable to other output formats, meaning that these changes will not be transferred to the HTML or e-book outputs.

12.4.3   E-books

The e-book formats can be read on devices like smartphones, tablets, or special e-readers such as Kindle. You can create an e-book of the EPUB format with `bookdown::epub_book`.

12.4.4   A single document

We highlighted in Section 12.3 that **bookdown** extends the syntax provided by R Markdown, allowing automatic numbering of figures / tables / equations, and cross-referencing them. You may use **bookdown** within single-file R Markdown documents to benefit from these features. The functions `html_document2()`, `tufte_html2()`, `pdf_document2()`, `word_document2()`, `tufte_handout2()`, and `tufte_book2()` are designed for this purpose. To use this in a traditional R Markdown document, you can replace the output YAML option as follows:

```
---
title: "Document Title"
output: bookdown::pdf_document2
---
```

12.5 Editing

In this section, we explain how to edit, build, preview, and serve the book locally.

12.5.1 Build the book

To build all Rmd files into a book, you can call the function `bookdown::render_book()`. It uses the settings specified in the `_output.yml` (if it exists). If multiple output formats are specified in it, all formats will be built. If you are using RStudio, this can be done through the `Build` tab. Open the drop down menu `Build Book` if you only want to build one format.

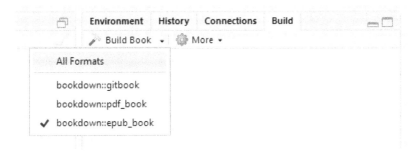

FIGURE 12.2: The `Build` tab within RStudio highlighting **bookdown** output formats.

12.5.2   Preview a chapter

Building the whole book can be slow when the size of the book is big or your book contains large amounts of computation. We can use the `preview_chapter()` function in **bookdown** to only build a single chapter at a time. Equivalently, you can click the `Knit` button in RStudio.

12.5.3   Serve the book

Instead of running `render_book()` or `preview_chapter()` each time you want to view the changes, you can use the function `bookdown::serve_book()` to start a live preview of the book. Any time a Rmd file is saved, the book will be recompiled automatically, and the preview will be updated to reflect the changes.

12.5.4   RStudio addins

The **bookdown** package comes with two addins for RStudio which assist the editing of books:

- "Preview Book": this calls `bookdown::serve_book()` to compile and serve the book.

- "Input LaTeX Math": provides a text box which allows you to write LaTeX equations, to avoid common errors when typing the raw LaTeX math expressions.

12.6   Publishing

You can generate books for both physical and electronic distribution. This section outlines some of the main options.

12.6.1 RStudio Connect

The easiest way to publish books online is through `https://bookdown.org`, which is a website provided by RStudio to host your books for free. Books can be pushed to this website by using `bookdown::publish_book()`. You will need to sign up for an account at `https://bookdown.org/connect/`, and your login details will be used to authorize **bookdown** the first time you call the `publish_book()` function.

12.6.2 Other services

You can host your book online with many other web services, such as Netlify or GitHub (via GitHub Pages[2]). Because the output from `bookdown::render_book()` is a collection of static files, you can host them using the same methods of hosting normal web pages.

12.6.3 Publishers

You can consider publishing physical copies of your book with a publisher or using self-publishing. Many publishers provide LaTeX style classes that can be used to set the overall appearance of the book, and these can be used easily by setting the `documentclass` option in the YAML metadata of `index.Rmd`. Further customization of the appearance of the PDF book can be achieved by altering the LaTeX preamble via the `includes: in_header` option of `bookdown::pdf_book`.

---

[2]`https://pages.github.com`

13

Journals

Academic journals often have strict guidelines on the formatting for submitted articles. As of today, few journals directly support R Markdown submissions, but many support the LaTeX format. While you can convert R Markdown to LaTeX (see Section 3.3), different journals have different typesetting requirements and LaTeX styles, and it may be slow and frustrating for all authors who want to use R Markdown to figure out the technical details about how to properly convert a paper based on R Markdown to a LaTeX document that meets the journal requirements.

The **rticles** package (Allaire et al., 2018a) is designed to simplify the creation of documents that conform to submission standards. A suite of custom R Markdown templates for popular journals is provided by the package such as those shown in Figure 13.2.

Understanding of LaTeX is recommended, but not essential, to use this package. R Markdown templates may sometimes inevitably contain LaTeX code, but usually we can use the simpler Markdown and **knitr** syntax to produce elements like figures, tables, and math equations as explained in Chapter 2.

13.1   Get started

You can install and use **rticles** from CRAN as follows:

```
# Install from CRAN
install.packages("rticles")

# Or install development version from GitHub
devtools::install_github("rstudio/rticles")
```

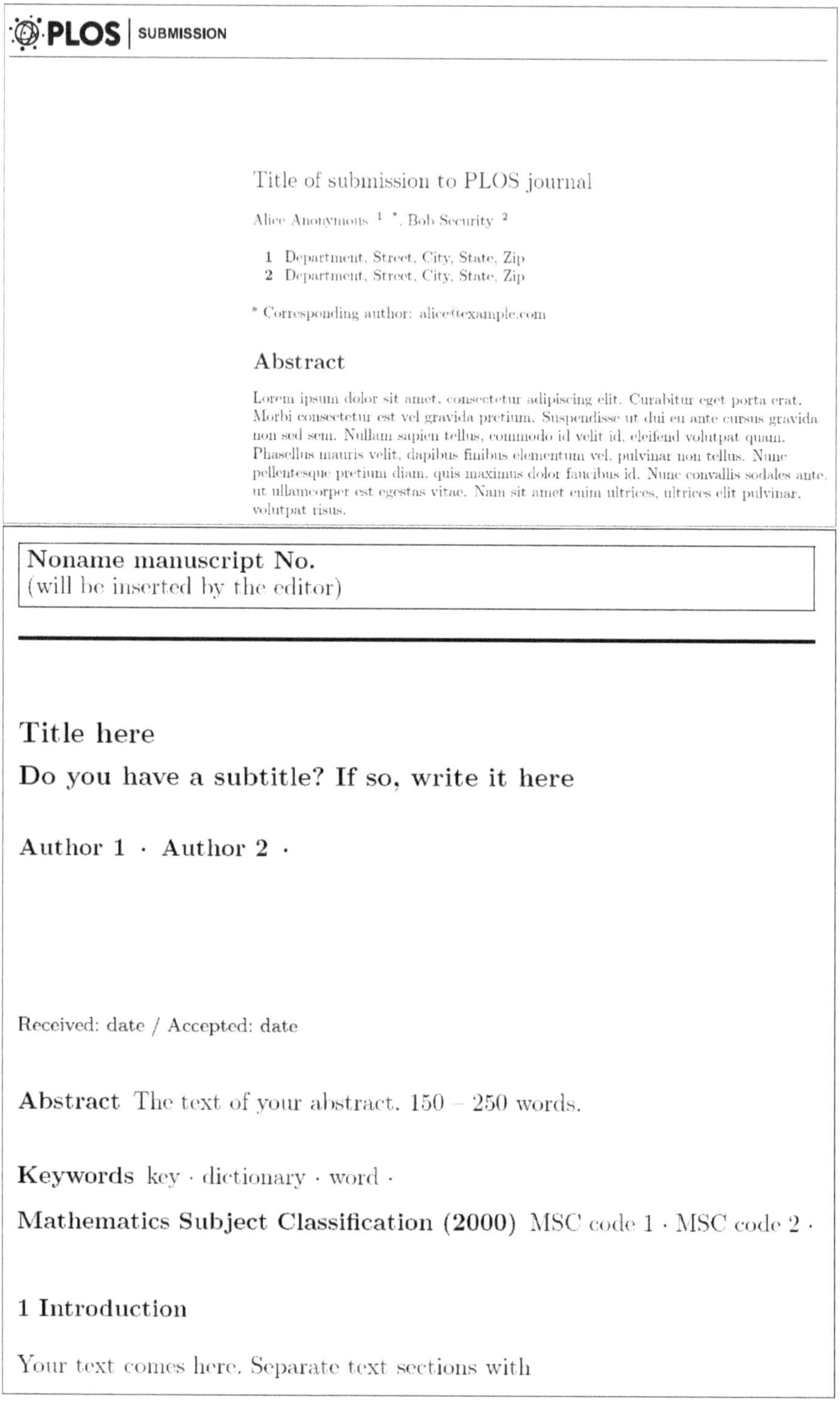

:◎: **PLOS** | SUBMISSION

Title of submission to PLOS journal

Alice Anonymous [1] [*], Bob Security [2]

1 Department, Street, City, State, Zip
2 Department, Street, City, State, Zip

* Corresponding author: alice@texample.com

Abstract

Lorem ipsum dolor sit amet, consectetur adipiscing elit. Curabitur eget porta erat. Morbi consectetur est vel gravida pretium. Suspendisse ut dui eu ante cursus gravida non sed sem. Nullam sapien tellus, commodo id velit id, eleifend volutpat quam. Phasellus mauris velit, dapibus finibus elementum vel, pulvinar non tellus. Nunc pellentesque pretium diam, quis maximus dolor faucibus id. Nunc convallis sodales ante, ut ullamcorper est egestas vitae. Nam sit amet enim ultrices, ultrices elit pulvinar, volutpat risus.

Noname manuscript No.
(will be inserted by the editor)

Title here
Do you have a subtitle? If so, write it here

Author 1 · Author 2 ·

Received: date / Accepted: date

Abstract The text of your abstract. 150 – 250 words.

Keywords key · dictionary · word ·

Mathematics Subject Classification (2000) MSC code 1 · MSC code 2 ·

1 Introduction

Your text comes here. Separate text sections with

FIGURE 13.1: Two journal templates in the **rticles** package (PLOS and Springer).

We would recommend the development version of the package from GitHub, as it contains the most up-to-date versions along with several new templates.

If you are using RStudio, you can easily access the templates through `File` `-> New File -> R Markdown`. This will open the dialog box where you can select from one of the available templates as shown in Figure 13.2.

FIGURE 13.2: The R Markdown template window in RStudio showing available **rticles** templates.

If you are using the command line, you can use the `rmarkdown::draft()` function, which requires you to specify a template, e.g.,

```
rmarkdown::draft(
  "MyJSSArticle.Rmd", template = "jss_article",
  package = "rticles"
)
```

13.2 rticles templates

The **rticles** package provides templates for various journals and publishers, including:

- JSS articles (Journal of Statistical Software)
- R Journal articles
- CTeX documents
- ACM articles (Association of Computing Machinery)
- ACS articles (American Chemical Society)
- AMS articles (American Meteorological Society)
- PeerJ articles
- Elsevier journal submissions
- AEA journal submissions (American Meteorological Society)
- IEEE Transaction journal submissions
- Statistics in Medicine journal submissions
- Royal Society Open Science journal submissions
- Bulletin de l'AMQ journal submissions
- MDPI journal submissions
- Springer journal submissions

The full list is available within the R Markdown templates window in RStudio, or through the command `getNamespaceExports("rticles")`.

13.3 Using a template

Templates have an extended YAML section compared to the basic R Markdown template, which allows you to specify additional details relevant to the custom template. Below is an example of the YAML section for the *Springer* template:

```
title: Title here
subtitle: Do you have a subtitle? If so, write it here
titlerunning: Short form of title (if too long for head)
authorrunning:
```

```
  Short form of author list if too long for running head
thanks: |
  Grants or other notes about the article that should go
  on the front page should be placed here. General
  acknowledgments should be placed at the end of the article.
authors:
  - name: Author 1
    address: Department of YYY, University of XXX
    email: abc@def
  - name: Author 2
    address: Department of ZZZ, University of WWW
    email: djf@wef
keywords:
  - key
  - dictionary
  - word
MSC:
  - MSC code 1
  - MSC code 2
abstract: |
  The text of your abstract.  150 -- 250 words.
bibliography: bibliography.bib
output: rticles::springer_article
```

As the Rmd documents are built using customized templates, you may not be able to use the YAML metadata to control the layout of the document as described in Section 3.3, unless the template supports such metadata. For example, adding `toc: true` may not add a table of contents. Commands that control the building process may still be used though, including `keep_tex: true`, or those that configure **knitr** chunk options (e.g., `fig_width`).

13.4 LaTeX content

As the only output format of the **rticles** formats is PDF, the content of the documents may include raw LaTeX formatting. This means you may use LaTeX to produce figures and tables (if you have to), e.g.,

```
\begin{figure}[ht]
\centering
\includegraphics[width=\linewidth]{foo}
\caption{An example image.}
\label{fig:foo}
\end{figure}
```

Unless you have specific requirements for using LaTeX, we recommend that you use the R Markdown syntax. This keeps you work generally more readable (in terms of the source document), and less prone to formatting errors. For example, the above code block would be better represented as:

```
```{r foo, out.width="100%", fig.cap="An example image."}
knitr::include_graphics("foo.png")
```
```

13.5 Linking with bookdown

As explained in Section 12.3, **bookdown** offers several extensions to the Markdown syntax, which can be particularly useful for academic writing, including cross-referencing of figures and tables. All **rticles** output formats are based on rmarkdown::pdf_document, and we can use them as the "base formats" for bookdown::pdf_document2, e.g.,

```
output:
  bookdown::pdf_document2:
    base_format: rticles::peerj_article
```

You can substitute `rticles::peerj_article` with the template you actually intend to use.

13.6 Contributing templates

If you take a look at the GitHub repository of **rticles** (`https://github.com/rstudio/rticles`), you will see that a lot of the templates have been contributed by the R community. If you are interested in improving them or adding more journal templates, you may want to read Chapter 17, which outlines how a template can be made for R Markdown. Basically these templates are defined to translate the Pandoc variables from the YAML frontmatter and the body of the R Markdown document into LaTeX.

14

Interactive Tutorials

The **learnr** package (Borges and Allaire, 2018) makes it easy to turn any R Markdown document into an interactive tutorial. Tutorials consist of content along with interactive components for checking and reinforcing understanding. Tutorials can include any or all of the following:

1. Narrative, figures, illustrations, and equations.

2. Code exercises (R code chunks that users can edit and execute directly).

3. Quiz questions.

4. Videos (currently supported services include YouTube and Vimeo).

5. Interactive Shiny components.

Tutorials automatically preserve work done within them, so if a user works on a few exercises or questions and returns to the tutorial later, they can pick up right where they left off.

This chapter is only a brief summary of **learnr**'s full documentation at `https://rstudio.github.io/learnr/`. If you are interested in building more sophisticated tutorials, we recommend that you read the full documentation.

14.1 Get started

To create a **learnr** tutorial, first install the **learnr** package with:

```
install.packages("learnr")
```

Then you can select the "Interactive Tutorial" template from the "New R Markdown" dialog in the RStudio IDE (see Figure 14.1).

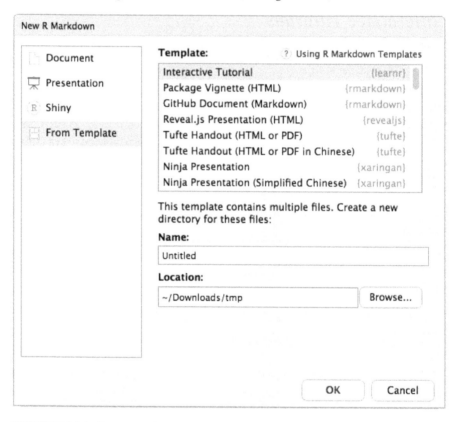

FIGURE 14.1: Create an interactive tutorial in RStudio.

If you do not use RStudio, it is also easy to create a tutorial: add `runtime: shiny_prerendered` and the output format `learnr::tutorial` to the YAML metadata, use `library(learnr)` within your Rmd file to activate the tutorial mode, and then add the chunk option `exercise = TRUE` to turn code chunks into exercises. Your tutorial users can edit and execute the R code and see the results right within their web browser.

Below is a minimal tutorial example:

```
---
title: "Hello, Tutorial!"
output: learnr::tutorial
runtime: shiny_prerendered
```

```
---

```{r setup, include=FALSE}
library(learnr)
```

This code computes the answer to one plus one, change it
so it computes two plus two:

```{r addition, exercise=TRUE}
1 + 1
```
```

To run this tutorial, you may hit the button "Run Document" in RStudio, or call the function rmarkdown::run() on this Rmd file. Figure 14.2 shows what the tutorial looks like in the browser. Users can do the exercise by editing the code and running it live in the browser.

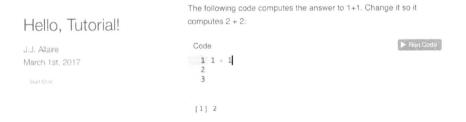

FIGURE 14.2: A simple example tutorial.

We strongly recommend that you assign unique chunk labels to exercises (e.g., the above example used the label addition), because chunk labels will be used as identifiers for **learnr** to save and restore user work. Without these identifiers, users could possibly lose their work in progress the next time when they pick up the tutorial.

14.2 Tutorial types

There are two main types of tutorial documents:

1. Tutorials that are mostly narrative and/or video content, and also include some runnable code chunks. These documents are very similar to package vignettes in that their principal goal is communicating concepts. The interactive tutorial features are then used to allow further experimentation by the reader.

2. Tutorials that provide a structured learning experience with multiple exercises, quiz questions, and tailored feedback.

The first type of tutorial is much easier to author while still being very useful. These documents will typically add `exercise` = `TRUE` to selected code chunks, and also set `exercise.eval` = `TRUE` so the chunk output is visible by default. The reader can simply look at the R code and move on, or play with it to reinforce their understanding.

The second type of tutorial provides much richer feedback and assessment, but also requires considerably more effort to author. If you are primarily interested in this sort of tutorial, there are many features in **learnr** to support it, including exercise hints and solutions, automated exercise checkers, and multiple choice quizzes with custom feedback.

The most straightforward path is to start with the first type of tutorial (executable chunks with pre-evaluated output), and then move into more sophisticated assessment and feedback over time.

14.3   Exercises

Exercises are interactive R code chunks that allow readers to directly execute R code and see its results. We have shown a simple exercise in Figure 14.2.

Exercises can include hints or solutions as well as custom checking code to provide feedback on user answers.

14.3.1   Solutions

To create a solution to an exercise in a code chunk with the chunk label `foo`, you add a new code chunk with the chunk label `foo-solution`, e.g.,

```{r filter, exercise=TRUE}
# Change the filter to select February rather than January
nycflights <- filter(nycflights, month == 1)
```

```{r filter-solution}
nycflights <- filter(nycflights, month == 2)
```

When a solution code chunk is provided, there will be a Solution button on the exercise (see Figure 14.3). Users can click this button to see the solution.

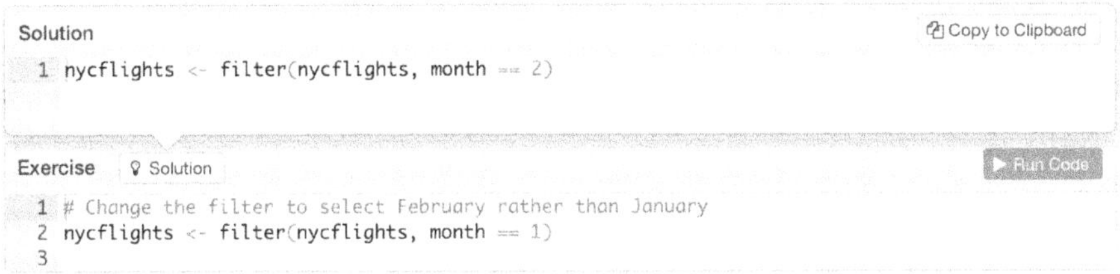

FIGURE 14.3: A solution to an exercise.

14.3.2 Hints

Sometimes you may not want to give the solutions directly to students, but provide hints instead to guide them. Hints can be either Markdown-based text content or code snippets.

To create a hint based on custom Markdown content, add a `<div>` tag with an id attribute that marks it as hint for your exercise (e.g., filter-hint). For example:

```{r filter, exercise=TRUE}
# filter the flights table to include only United and
# American flights
flights
```

<div id="filter-hint">
```
```

```
**Hint:** You may want to use the dplyr `filter` function.
</div>
```

The content within the `<div>` will be displayed underneath the R code editor for the exercise whenever the user presses the `Hint` button.

If your Pandoc version is higher than 2.0 (check `rmark-down::pandoc_version()`), you can also use the alternative syntax to write the `<div>`:

```
:::{#filter-hint}
**Hint:** You may want to use the dplyr `filter` function.
:::
```

To create a hint with a code snippet, you add a new code chunk with the label suffix -hint, e.g.,

````
```{r filter, exercise=TRUE}
# filter the flights table to include only United and
# American flights
flights
```
````

````
```{r filter-hint}
filter(flights, ...)
```
````

You can also provide a sequence of hints that reveal progressively more of the solution as desired by the user. To do this, create a sequence of indexed hint chunks (e.g., -hint-1, -hint-2, -hint-3, etc.) for your exercise chunk. For example:

````
```{r filter, exercise=TRUE}
# filter the flights table to include only United and
# American flights
flights
```
````

````
```{r filter-hint-1}
````
````

```
filter(flights, ...)
```

```
```{r filter-hint-2}
filter(flights, UniqueCarrier == "AA")
```
```

```
```{r filter-hint-3}
filter(flights, UniqueCarrier == "AA" | UniqueCarrier == "UA")
```
```

14.4 Quiz questions

You can include one or more multiple-choice quiz questions within a tutorial to help verify that readers understand the concepts presented. Questions can either have a single or multiple correct answers.

Include a question by calling the question() function within an R code chunk, e.g.,

```{r letter-a, echo=FALSE}
question("What number is the letter A in the English alphabet?",
  answer("8"),
  answer("14"),
  answer("1", correct = TRUE),
  answer("23")
)
```

Figure 14.4 shows what the above question would look like within a tutorial.

The functions question() and answer() have several other arguments for more features that allow you to customize the questions and answers, such as custom error messages when the user's answer is wrong, allowing users to retry a question, multiple-choice questions, and multiple questions in a group. See their help pages in R for more information.

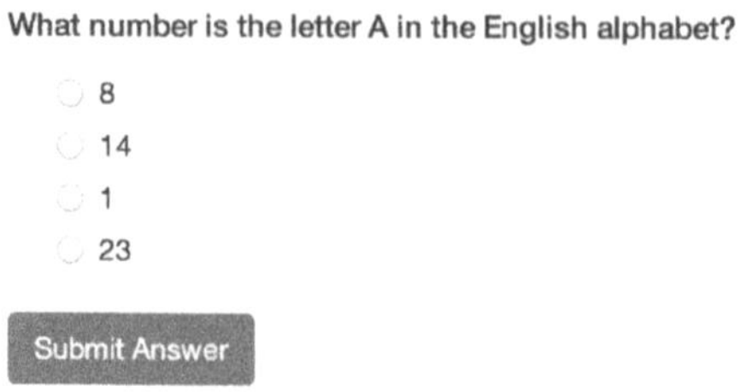

What number is the letter A in the English alphabet?

- ○ 8
- ○ 14
- ○ 1
- ○ 23

[Submit Answer]

FIGURE 14.4: A question in a tutorial.

14.5 Videos

You can include videos published on either YouTube or Vimeo within a tutorial using the standard Markdown image syntax. Note that any valid YouTube or Vimeo URL will work. For example, the following are all valid examples of video embedding:

```
![](https://youtu.be/zNzZ1PfUDNk)
![](https://www.youtube.com/watch?v=zNzZ1PfUDNk)

![](https://vimeo.com/142172484)
![](https://player.vimeo.com/video/142172484)
```

Videos are responsively displayed at 100% of their container's width (with height automatically determined based on a 16x9 aspect ratio). You can change this behavior by adding attributes to the Markdown code where you reference the video.

You can specify an alternate percentage for the video's width or an alternate fixed width and height. For example:

```
![](https://youtu.be/zNzZ1PfUDNk){width="90%"}

![](https://youtu.be/zNzZ1PfUDNk){width="560" height="315"}
```

14.6 Shiny components

Tutorials are essentially Shiny documents, which we will introduce in Chapter 19. For that reason, you are free to use any interactive Shiny components in tutorials, not limited to exercises and quiz questions.

The Shiny UI components can be written in normal R code chunks. For the Shiny server logic code (rendering output), you need to add a chunk option `context="server"` to code chunks. For example:

````
```{r, echo=FALSE}
sliderInput("bins", "Number of bins:", 30, min = 1, max = 50)
plotOutput("distPlot")
```
````

````
```{r, context="server"}
output$distPlot = renderPlot({
  x = faithful[, 2]  # Old Faithful Geyser data
  bins = seq(min(x), max(x), length.out = input$bins + 1)
  hist(x, breaks = bins, col = 'darkgray', border = 'white')
})
```
````

Again, since tutorials are Shiny applications, they can be deployed using the same methods mentioned in Section 19.2.

14.7 Navigation and progress tracking

Each **learnr** tutorial includes a table of contents on the left that tracks student progress (see Figure 14.5). Your browser will remember which sections of a tutorial a student has completed, and return a student to where he/she left off when the tutorial is reopened.

You can optionally reveal content by one sub-section at a time. You can use

FIGURE 14.5: Keeping track of the student's progress in a tutorial.

this feature to let students set their own pace, or to hide information that would spoil an exercise or question that appears just before it.

To use progressive reveal, set the `progressive` option to `true` in the `learnr::tutorial` output format in the YAML metadata, e.g.,

```
---
title: "Programming basics"
output:
  learnr::tutorial:
    progressive: true
    allow_skip: true
runtime: shiny_prerendered
---
```

The `allow_skip` option above indicates that students can skip any sections, and move directly to the next section without completing exercises in the previous section.

Part IV

Other Topics

15

Parameterized reports

One of the many benefits of working with R Markdown is that you can reproduce analysis at the click of a button. This makes it very easy to update any work and alter any input parameters within the report. Parameterized reports extend this one step further, and allow users to specify one or more parameters to customize the analysis. This is useful if you want to create a report template that can be reused across multiple similar scenarios. Examples may include:

- Showing results for a specific geographic location.

- Running a report that covers a specific time period.

- Running a single analysis multiple times for different assumptions.

- Controlling the behavior of **knitr** (e.g., specify if you want the code to be displayed or not).

In this chapter, we discuss the use of parameterized reports, and explain how we can interactively define the parameters to compile the results.

15.1 Declaring parameters

Parameters are specified using the params field within the YAML section. We can specify one or more parameters with each item on a new line. As an example:

```
---
title: My Document
output: html_document
params:
```

```
year: 2018
region: Europe
printcode: TRUE
data: file.csv
---
```

All standard R types that can be parsed by `yaml::yaml.load()` can be included as parameters, including `character`, `numeric`, `integer`, and `logical` types. We can also use R objects by including `!r` before R expressions. For example, we could include the current date with the following R code:

```
---
title: My Document
output: html_document
params:
  date: !r Sys.Date()
---
```

Any R expressions included within the parameters are executed before any code in the document, therefore any package dependencies must be explicitly stated using the `package::function` notation (e.g., `!r lubridate::today()`), even if the package is loaded later in the Rmd document.

15.2   Using parameters

You can access the parameters within the knitting environment and the R console in RStudio.[1] The values are contained within a read-only list called `params`. In the previous example, the parameters can be accessed as follows:

```
params$year
params$region
```

Parameters can also be used to control the behavior of **knitr**. For example,

---

[1]Parameters will not be available immediately after loading the file, but require any line of the report to be executed first.

the **knitr** chunk option echo controls whether to display the program code, and we can set this option globally in a document via a parameter:

```
---
params:
  printcode: false  # or set it to true
---
```

```
```{r, setup, include=FALSE}
# set this option in the first code chunk in the document
knitr::opts_chunk$set(echo = params$printcode)
```
```

15.3   Knitting with parameters

There are three ways in which a parameterized report can be knitted:

- Using the Knit button within RStudio.
- rmarkdown::render() with the params argument.
- Using an interactive user interface to input parameter values.

15.3.1   The Knit button

By using the Knit button in RStudio or calling rmarkdown::render() function, the default values listed in the YAML metadata (if specified) will be used.

15.3.2   Knit with custom parameters

Even if your document has the params field in the YAML metadata, you can actually override it by providing a custom list of parameter values to the function rmarkdown::render(). For example:

```
rmarkdown::render("MyDocument.Rmd", params = list(
  year = 2017,
  region = "Asia",
  printcode = FALSE,
  file = "file2.csv"
))
```

We do not have to explicitly state all parameters in the params argument. Any parameters not specified will default to the values specified in the YAML metadata. For example, this will only override the region parameter:

```
rmarkdown::render("MyDocument.Rmd", params = list(
  region = "Asia"
))
```

You may want to integrate these changes into a function. Such a function could also be used to create an output file with a different filename for each of the different combination of parameters. In the following example, a new file Report-region-year.pdf is created for each set of parameters:

```
render_report = function(region, year) {
  rmarkdown::render(
    "MyDocument.Rmd", params = list(
      region = region,
      year = year
    ),
    output_file = paste0("Report-", region, "-", year, ".pdf")
  )
}
```

15.3.3   The interactive user interface

We can use a graphical user interface (GUI) based on Shiny to interactively input the parameters of a report. The user interface can be called by either rmarkdown::render("MyDocument.Rmd", params = "ask") or clicking the drop-down menu behind the Knit button and choosing Knit with Parameters in RStudio. Figure 15.1 shows the GUI of **rmarkdown** asking for inputting parameters.

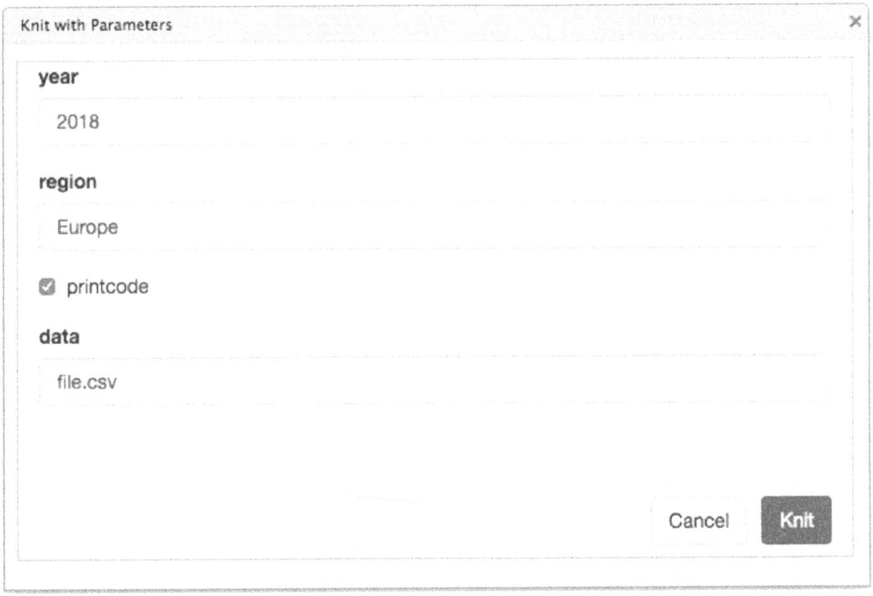

FIGURE 15.1: Input parameter values interactively for parameterized reports.

The input controls for different types of parameters can be customized by specifying additional sub-items within the parameter specification in YAML. For example, sliders, check boxes, and text input boxes can all be used for input controls.

In addition, we can also specify constraints of the values allowed in each parameter. For example, we may only want our model to be run for years between 2010 and 2018. This is particularly beneficial if you would like other users to interact with the report, as it prevents users from attempting to run reports outside of the designed limits.

Adapting our above example to include some settings:

```
---
title: My Document
output: html_document
params:
  year:
```

```
    label: "Year"
    value: 2017
    input: slider
    min: 2010
    max: 2018
    step: 1
    sep: ""
  region:
    label: "Region:"
    value: Europe
    input: select
    choices: [North America, Europe, Asia, Africa]
  printcode:
    label: "Display Code:"
    value: TRUE
  data:
    label: "Input dataset:"
    value: results.csv
    input: file
---
```

This results in the user interface for the parameters as shown in Figure 15.2.

The type of Shiny control used is controlled by the `input` field. Table 15.1 shows the input types currently supported (see the help page for the associated Shiny function for additional attributes that can be specified to customize the input, e.g., `?shiny::checkboxInput`).

15.4 Publishing

Parameterized reports are supported by the publishing platform RStudio Connect (`https://www.rstudio.com/products/connect/`). If you publish a parameterized report to an RStudio Connect server, you will be able to compile reports by interactively choosing different parameter values on the server, and easily store/navigate through different reports built previously. You may watch a video demonstration at `https://bit.ly/rsc-params`.

FIGURE 15.2: Custom controls for parameters.

TABLE 15.1: Possible input types and the associated Shiny functions for parameterized reports.

Input Type	Shiny Function
checkbox	checkboxInput
numeric	numericInput
slider	sliderInput
date	dateInput
text	textInput
file	fileInput
radio	radioButtons
select	selectInput
password	passwordInput

16

HTML Widgets

We briefly mentioned HTML widgets in the beginning of this book in Section 2.8.1. The **htmlwidgets** package (Vaidyanathan et al., 2018) provides a framework for creating R bindings to JavaScript libraries. HTML Widgets can be:

- Used at the R console for data analysis just like conventional R plots.

- Embedded within R Markdown documents.[1]

- Incorporated into Shiny web applications.

- Saved as standalone web pages for ad-hoc sharing via Email and Dropbox, etc.

There have been many R packages developed based on the HTML widgets framework, to make it easy for R users to create JavaScript applications using pure R syntax and data. It is not possible to introduce all these R packages in this chapter. Readers should read the documentation of specific widget packages for the usage. This chapter is mainly for developers who want to bring more JavaScript libraries into R, and it requires reasonable familiarity with the JavaScript language.

16.1 Overview

By following a small set of conventions, it is possible to create HTML widgets with very little code. All widgets include the following components:

---

[1]Note that interactivity only works when the output format is HTML, including HTML documents and presentations. If the output format is not HTML, it is possible to automatically create and embed a static screenshot of the widget instead. See Section 2.8.1 for more information.

1. **Dependencies**. These are the JavaScript and CSS assets used by the widget (e.g., the library for which you are creating a wrapper).

2. **R binding**. This is the function that end-users will call to provide input data to the widget and specify various options for how the widget should render. This also includes some short boilerplate functions required to use the widget within Shiny applications.

3. **JavaScript binding**. This is the JavaScript code that glues everything together, passing the data and options gathered in the R binding to the underlying JavaScript library.

HTML widgets are always hosted within an R package, and should include all of the source code for their dependencies. This is to ensure that R code that renders widgets is fully reproducible (i.e., it does not require an Internet connection or the ongoing availability of an Internet service to run).

16.2 A widget example (sigma.js)

To start with, we will walk through the creation of a simple widget that wraps the sigma.js[2] graph visualization library. When we are done, we will be able to use it to display interactive visualizations of GEXF[3] (Graph Exchange XML Format) data files. For example (see Figure 16.1 for the output, which is interactive if you are reading the HTML version of this book):

```
library(sigma)
d = system.file("examples/ediaspora.gexf.xml", package = "sigma")
sigma(d)
```

There is remarkably little code required to create this binding. Next we will go through all of the components step by step. Then we will describe how you can create your own widgets, including automatically generating basic scaffolding for all of the core components.

---

[2]http://sigmajs.org
[3]http://gexf.net

FIGURE 16.1: A graph generated using the sigma.js library and the sigma package.

16.2.1 File layout

Let's assume that our widget is named **sigma** and is located within an R package of the same name. Our JavaScript binding source code file is named sigma.js. Since our widget will read GEXF data files, we will also need to include both the base `sigma.min.js` library and its GEXF plugin. Here are the files that we will add to the package:

```
R/
| sigma.R

inst/
|-- htmlwidgets/
|   |-- sigma.js
|   |-- sigma.yaml
|   |-- lib/
|   |   |-- sigma-1.0.3/
|   |   |   |-- sigma.min.js
|   |   |   |-- plugins/
|   |   |   |   |-- sigma.parsers.gexf.min.js
```

Note the convention that the JavaScript, YAML, and other dependencies are all contained within the `inst/htmlwidgets` directory, which will subsequently be installed into a package sub-directory named `htmlwidgets`.

16.2.2 Dependencies

Dependencies are the JavaScript and CSS assets used by a widget, included within the `inst/htmlwidgets/lib` directory. They are specified using a YAML configuration file that uses the name of the widget as its base filename. Here is what our **sigma.yaml** file looks like:

```
dependencies:
  - name: sigma
    version: 1.0.3
    src: htmlwidgets/lib/sigma-1.0.3
    script:
      - sigma.min.js
      - plugins/sigma.parsers.gexf.min.js
```

The dependency `src` specification refers to the directory that contains the library, and `script` refers to specific JavaScript files. If your library contains multiple JavaScript files specify each one on a line beginning with - as shown above. You can also add `stylesheet` entries, and even `meta` or `head` entries. Multiple dependencies may be specified in one YAML file. See the documentation on the `htmlDependency()` function in the **htmltools** package for additional details.

16.2.3 R binding

We need to provide users with an R function that invokes our widget. Typically this function will accept input data as well as various options that control the widget's display. Here is the R function for the `sigma` widget:

```
#' @import htmlwidgets
#' @export
sigma = function(
  gexf, drawEdges = TRUE, drawNodes = TRUE, width = NULL,
```

```
  height = NULL
) {

  # read the gexf file
  data = paste(readLines(gexf), collapse = "\n")

  # create a list that contains the settings
  settings = list(drawEdges = drawEdges, drawNodes = drawNodes)

  # pass the data and settings using 'x'
  x = list(data = data, settings = settings)

  # create the widget
  htmlwidgets::createWidget(
    "sigma", x, width = width, height = height
  )
}
```

The function takes two classes of input: the GEXF data file to render, and some additional settings that control how it is rendered. This input is collected into a list named x, which is then passed on to the htmlwidgets::createWidget() function. This x variable will subsequently be made available to the JavaScript binding for sigma (to be described in the next section). Any width or height parameter specified is also forwarded to the widget (widgets size themselves automatically by default, so typically do not require an explicit width or height).

We want our sigma widget to also work in Shiny applications, so we add the following boilerplate Shiny output and render functions (these are always the same for all widgets):

```
#' @export
sigmaOutput = function(outputId, width = "100%", height = "400px") {
  htmlwidgets::shinyWidgetOutput(
    outputId, "sigma", width, height, package = "sigma"
  )
}
#' @export
renderSigma = function(expr, env = parent.frame(), quoted = FALSE) {
```

```r
  if (!quoted) { expr = substitute(expr) } # force quoted
  htmlwidgets::shinyRenderWidget(
    expr, sigmaOutput, env, quoted = TRUE
  )
}
```

16.2.4   JavaScript binding

The third piece in the puzzle is the JavaScript required to activate the widget. By convention, we will define our JavaScript binding in the file inst/htmlwidgets/sigma.js. Here is the full source code of the binding:

```javascript
HTMLWidgets.widget({

  name: "sigma",

  type: "output",

  factory: function(el, width, height) {

    // create our sigma object and bind it to the element
    var sig = new sigma(el.id);

    return {
      renderValue: function(x) {

        // parse gexf data
        var parser = new DOMParser();
        var data = parser.parseFromString(x.data, "application/xml");

        // apply settings
        for (var name in x.settings)
          sig.settings(name, x.settings[name]);

        // update the sigma object
        sigma.parsers.gexf(
          data,            // parsed gexf data
```

```
    sig,            // sigma object
    function() {
      // need to call refresh to reflect new settings
      // and data
      sig.refresh();
    }
  );
},

resize: function(width, height) {

  // forward resize on to sigma renderers
  for (var name in sig.renderers)
    sig.renderers[name].resize(width, height);
},

// make the sigma object available as a property on the
// widget instance we are returning from factory(). This
// is generally a good idea for extensibility -- it helps
// users of this widget interact directly with sigma,
// if needed.
s: sig
};
}
});
```

We provide a name and type for the widget, plus a factory function that takes el (the HTML element that will host this widget), width, and height (width and height of the HTML element, in pixels — you can always use offsetWidth and offsetHeight for this).

The factory function should prepare the HTML element to start receiving values. In this case, we create a new sigma element and pass it to the id of the DOM element that hosts the widget on the page.

We are going to need access to the sigma object later (to update its data and settings), so we save it as a variable sig. Note that variables declared directly inside of the factory function are tied to a particular widget instance (el).

The return value of the factory function is called a *widget instance object*. It is a bridge between the htmlwidgets runtime, and the JavaScript visualization that you are wrapping. As the name implies, each widget instance object is responsible for managing a single widget instance on a page.

The widget instance object you create must have one required method, and may have one optional method:

1. The required `renderValue` method actually pours our dynamic data and settings into the widget's DOM element. The x parameter contains the widget data and settings. We parse and update the GEXF data, apply the settings to our previously-created `sig` object, and finally call `refresh` to reflect the new values on-screen. This method may be called repeatedly with different data (i.e., in Shiny), so be sure to account for that possibility. If it makes sense for your widget, consider making your visualization transition smoothly from one value of x to another.

2. The optional `resize` method is called whenever the element containing the widget is resized. The only reason not to implement this method is if your widget naturally scales (without additional JavaScript code needing to be invoked) when its element size changes. In the case of sigma.js, we forward the sizing information on to each of the underlying sigma renderers.

All JavaScript libraries handle initialization, binding to DOM elements, dynamically updating data, and resizing slightly differently. Most of the work on the JavaScript side of creating widgets is mapping these three functions, `factory`, `renderValue`, and `resize`, correctly onto the behavior of the underlying library.

The sigma.js example uses a simple object literal to create its widget instance object, but you can also use class based objects[4] or any other style of object, as long as `obj.renderValue(x)` and `obj.resize(width, height)` can be invoked on it.

You can add additional methods and properties on the widget instance object. Although they will not be called by htmlwidgets itself, they might be useful to users of your widget that know some JavaScript and want to further customize your widget by adding custom JS code (e.g., using the R function

---

[4]`https://developer.mozilla.org/en-US/docs/Web/JavaScript/Reference/Classes`

`htmlwidgets::onRender()`). In this case, we add a property s to make the sigma object itself available.

```
library(sigma)
library(htmlwidgets)
library(magrittr)
d = system.file("examples/ediaspora.gexf.xml", package = "sigma")
sigma(d) %>% onRender("function(el, x) {
  // this.s is the sigma object
  console.log(this.s);
}")
```

16.2.5   Demo

Our widget is now complete! If you want to test drive it without reproducing all of the code locally you can install it from GitHub as follows:

```
devtools::install_github('jjallaire/sigma')
```

Here is the code to try it out with some sample data included with the package:

```
library(sigma)
sigma(system.file("examples/ediaspora.gexf.xml", package = "sigma"))
```

If you execute this code in the R console, you will see the widget displayed in the RStudio Viewer (or in an external browser if you are not running RStudio). If you include it within an R Markdown document, the widget will be embedded into the document.

We can also use the widget in a Shiny application:

```
library(shiny)
library(sigma)

gexf = system.file("examples/ediaspora.gexf.xml", package = "sigma")

ui = shinyUI(fluidPage(
```

```r
  checkboxInput("drawEdges", "Draw Edges", value = TRUE),
  checkboxInput("drawNodes", "Draw Nodes", value = TRUE),
  sigmaOutput('sigma')
))

server = function(input, output) {
  output$sigma = renderSigma(
    sigma(gexf,
          drawEdges = input$drawEdges,
          drawNodes = input$drawNodes)
  )
}

shinyApp(ui = ui, server = server)
```

16.3   Creating your own widgets

16.3.1   Requirements

To implement a widget, you can create a new R package that in turn depends on the **htmlwidgets** package. You can install the package from CRAN as follows:

```r
install.packages("htmlwidgets")
```

While it is not strictly required, the step-by-step instructions below for getting started also make use of the **devtools** package, which you can also install from CRAN:

```r
install.packages("devtools")
```

It is also possible to implement a widget without creating an R package, but it requires you to understand more about HTML dependencies (htmltools::htmlDependency()). We have given an example in Section 16.5.

16.3.2 Scaffolding

To create a new widget, you can call the `scaffoldWidget()` function to generate the basic structure for your widget. This function will:

- Create the `.R`, `.js`, and `.yaml` files required for your widget;
- If provided, take a Bower[5] package name and automatically download the JavaScript library (and its dependencies) and add the required entries to the `.yaml` file.

This method is highly recommended, as it ensures that you get started with the right file structure. Here is an example that assumes you want to create a widget named 'mywidget' in a new package of the same name:

```
# create package using devtools
devtools::create("mywidget")

# navigate to package dir
setwd("mywidget")

# create widget scaffolding
htmlwidgets::scaffoldWidget("mywidget")

# install the package so we can try it
devtools::install()
```

This creates a simple widget that takes a single `text` argument and displays that text within the widgets HTML element. You can try it like this:

```
library(mywidget)
mywidget("hello, world")
```

This is the most minimal widget possible, and does not yet include a JavaScript library to interface to (note that `scaffoldWidget()` can optionally include JavaScript library dependencies via the `bowerPkg` argument). Before getting started with development, you should review the introductory example above to make sure you understand the various components, and also review the additional articles and examples linked to in the next section.

---

[5]`https://bower.io`

16.3.3 Other packages

Studying the source code of other packages is a great way to learn more about creating widgets:

1. The **networkD3**[6] package illustrates creating a widget on top of D3[7], using a custom sizing policy for a larger widget, and providing multiple widgets from a single package.

2. The **dygraphs**[8] package illustrates using widget instance data, handling dynamic re-sizing, and using **magrittr** to decompose a large and flat JavaScript API into a more modular and pipeable R API.

3. The **sparkline**[9] package illustrates providing a custom HTML generation function (since sparklines must be housed in `<span>` rather than `<div>` elements).

If you have questions about developing widgets or run into problems during development, please do not hesitate to post an issue on the project's GitHub repository: `https://github.com/ramnathv/htmlwidgets/issues`.

16.4 Widget sizing

In the spirit of HTML widgets working just like plots in R, it is important that HTML widgets intelligently size themselves to their container, be it the RStudio Viewer, a figure in a **knitr** document, or a UI panel within a Shiny application. The **htmlwidgets** framework provides a rich mechanism for specifying the sizing behavior of widgets.

This sizing mechanism is designed to address the following constraints that affect the natural size of a widget:

*   **The kind of widget it is.** Some widgets may only be designed to look good at small, fixed sizes (like sparklines) while other widgets may want every pixel that can be spared (like network graphs).

---

[6]`https://github.com/christophergandrud/networkD3`
[7]`http://d3js.org`
[8]`https://github.com/rstudio/dygraphs/`
[9]`https://github.com/htmlwidgets/sparkline`

- **The context into which the widget is rendered.** While a given widget might look great at 960px by 480px in an R Markdown document, the same widget would look silly at that size in the RStudio Viewer pane, which is typically much smaller.

Widget sizing is handled in two steps:

1. First, a sizing policy is specified for the widget. This is done via the `sizingPolicy` argument to the `createWidget` function. Most widgets can accept the default sizing policy (or override only one or two aspects of it) and get satisfactory sizing behavior (see details below).

2. The sizing policy is used by the framework to compute the correct width and height for a widget given where it is being rendered. This size information is then passed to the `factory` and `resize` methods of the widget's JavaScript binding. It is up to the widget to forward this size information to the underlying JavaScript library.

16.4.1 Specifying a sizing policy

The default HTML widget sizing policy treats the widget with the same sizing semantics as an R plot. When printed at the R console, the widget is displayed within the RStudio Viewer and sized to fill the Viewer pane (modulo any padding). When rendered inside an R Markdown document, the widget is sized based on the default size of figures in the document.

Note that for most widgets the default sizing behavior is fine, and you will not need to create a custom sizing policy. If you need a slightly different behavior than the default, you can also selectively override the default behavior by calling the `sizingPolicy()` function and passing the result to `createWidget()`. For example:

```
htmlwidgets::createWidget(
  "sigma",
  x,
  width = width,
  height = height,
  sizingPolicy = htmlwidgets::sizingPolicy(
    viewer.padding = 0,
```

```
    viewer.paneHeight = 500,
    browser.fill = TRUE
  )
)
```

Below are two examples:

- The **networkD3** package uses custom sizing policies for all of its widgets. The simpleNetwork widget eliminates padding (as D3.js is already providing padding), and specifies that it wants to fill up as much space as possible when displayed in a standalone web browser:

```
sizingPolicy(padding = 0, browser.fill = TRUE)
```

- The sankeyNetwork widget requires much more space than is afforded by the RStudio Viewer or a typical **knitr** figure, so it disables those automatic sizing behaviors. It also provides a more reasonable default width and height for **knitr** documents:

```
sizingPolicy(viewer.suppress = TRUE,
             knitr.figure = FALSE,
             browser.fill = TRUE,
             browser.padding = 75,
             knitr.defaultWidth = 800,
             knitr.defaultHeight = 500)
```

Table 16.1 shows the various options that can be specified within a sizing policy. Note that the default width, height, and padding will be overridden if their values for a specific viewing context are provided (e.g., browser.defaultWidth will override defaultWidth when the widget is viewed in a web browser). Also note that when you want a widget a fill a viewer, the padding is still applied.

16.4.2 JavaScript resize method

Specifying a sizing policy allows **htmlwidgets** to calculate the width and height of your widget based on where it is being displayed. However, you

TABLE 16.1: Options that can be specified within a sizing policy.

Option	Description
defaultWidth	Default widget width in all contexts (browser, viewer, and knitr).
defaultHeight	Similar to `defaultWidth`, but for heights instead.
padding	The padding (in pixels) in all contexts.
viewer.defaultWidth	Default widget width within the RStudio Viewer.
viewer.defaultHeight	Similar to `viewer.defaultWidth`.
viewer.padding	Padding around the widget in the RStudio Viewer (defaults to 15 pixels).
viewer.fill	When displayed in the RStudio Viewer, automatically size the widget to the viewer dimensions. Default to `TRUE`.
viewer.suppress	Never display the widget within the RStudio Viewer (useful for widgets that require a large amount of space for rendering). Defaults to `FALSE`.
viewer.paneHeight	Request that the RStudio Viewer be forced to a specific height when displaying this widget.
browser.defaultWidth	Default widget width within a standalone web browser.
browser.defaultHeight	Similar to `browser.defaultWidth`.
browser.padding	Padding in a standalone browser (defaults to 40 pixels).
browser.fill	When displayed in a standalone web browser, automatically size the widget to the browser dimensions. Defaults to `FALSE`.
browser.external	Always use an external browser (via `browseURL()`). Defaults to `FALSE`, which will result in the use of an internal browser within RStudio v1.1 and higher.
knitr.defaultWidth	Default widget width within documents generated by **knitr** (e.g., R Markdown).
knitr.defaultHeight	Similar to `knitr.defaultWidth`.
knitr.figure	Apply the default **knitr** `fig.width` and `fig.height` to the widget rendered in R Markdown. Defaults to `TRUE`.

still need to forward this sizing information on to the underlying JavaScript library for your widget.

Every JavaScript library handles dynamic sizing a bit differently. Some do it automatically, some have a resize() call to force a layout, and some require that size be set only along with data and other options. Whatever the case it is, the **htmlwidgets** framework will pass the computed sizes to both your factory function and resize function. Here is a sketch of a JavaScript binding:

```
HTMLWidgets.widget({

  name: "demo",

  type: "output",

  factory: function(el, width, height) {

    return {
      renderValue: function(x) {

      },

      resize: function(width, height) {

      }
    };
  }
});
```

What you do with the passed width and height is up to you, and depends on the re-sizing semantics of the underlying JavaScript library. A couple of illustrative examples are included below:

- In the dygraphs widget (https://rstudio.github.io/dygraphs), the implementation of re-sizing is relatively simple, since the **dygraphs** library includes a resize() method to automatically size the graph to its enclosing HTML element:

```
resize: function(width, height) {
  if (dygraph)
    dygraph.resize();
}
```

- In the `forceNetwork` widget (https://christophergandrud.github.io/ networkD3/#force), the passed width and height are applied to the `<svg>` element that hosts the D3 network visualization, as well as forwarded on to the underlying D3 force simulation object:

```
factory: function(el, width, height) {

  var force = d3.layout.force();

  d3.select(el).append("svg")
    .attr("width", width)
    .attr("height", height);

  return {
    renderValue: function(x) {
      // implementation excluded
    },

    resize: function(width, height) {

      d3.select(el).select("svg")
        .attr("width", width)
        .attr("height", height);

      force.size([width, height]).resume();
    }
  };
}
```

As you can see, re-sizing is handled in a wide variety of fashions in different JavaScript libraries. The `resize` method is intended to provide a flexible way to map the automatic sizing logic of **htmlwidgets** directly into the underlying library.

16.5 Advanced topics

This section covers several aspects of creating widgets that are not required by all widgets, but are an essential part of getting bindings to certain types of JavaScript libraries to work properly. Topics covered include:

- Transforming JSON representations of R objects into representations required by JavaScript libraries (e.g., an R data frame to a D3 dataset).

- Passing JavaScript functions from R to JavaScript (e.g., a user-provided formatting or drawing function)

- Generating custom HTML to enclose a widget (the default is a `<div>`, but some libraries require a different element, e.g., a `<span>`).

- Creating a widget without creating an R package in the first place.

16.5.1 Data transformation

R objects passed as part of the `x` parameter to the `createWidget()` function are transformed to JSON using the internal function `htmlwidgets:::toJSON()`[10], which is basically a wrapper function of `jsonlite::toJSON()` by default. However, sometimes this representation is not what is required by the JavaScript library you are interfacing with. There are two JavaScript functions that you can use to transform the JSON data.

16.5.1.1 HTMLWidgets.dataframeToD3()

R data frames are represented in "long" form (an array of named vectors) whereas D3 typically requires "wide" form (an array of objects each of which includes all names and values). Since the R representation is smaller in size and much faster to transmit over the network, we create the long-form representation of R data, and then transform the data in JavaScript using the `dataframeToD3()` helper function.

Here is an example of the long-form representation of an R data frame:

---

[10]Note that it is not exported from **htmlwidgets**, so you are not supposed to call this function directly.

```
{
  "Sepal.Length": [5.1, 4.9, 4.7],
  "Sepal.Width": [3.5, 3, 3.2],
  "Petal.Length": [1.4, 1.4, 1.3],
  "Petal.Width": [0.2, 0.2, 0.2],
  "Species": ["setosa", "setosa", "setosa"]
}
```

After we apply `HTMLWidgets.dataframeToD3()`, it will become:

```
[
  {
    "Sepal.Length": 5.1,
    "Sepal.Width": 3.5,
    "Petal.Length": 1.4,
    "Petal.Width": 0.2,
    "Species": "setosa"
  },
  {
    "Sepal.Length": 4.9,
    "Sepal.Width": 3,
    "Petal.Length": 1.4,
    "Petal.Width": 0.2,
    "Species": "setosa"
  },
  {
    "Sepal.Length": 4.7,
    "Sepal.Width": 3.2,
    "Petal.Length": 1.3,
    "Petal.Width": 0.2,
    "Species": "setosa"
  }
]
```

As a real example, the `simpleNetwork` (https://christophergandrud. github.io/networkD3/#simple) widget accepts a data frame containing network links on the R side, and transforms it to a D3 representation within the JavaScript `renderValue` function:

```
renderValue: function(x) {

  // convert links data frame to d3 friendly format
  var links = HTMLWidgets.dataframeToD3(x.links);

  // ... use the links, etc ...

}
```

16.5.1.2 HTMLWidgets.transposeArray2D()

Sometimes a 2-dimensional array requires a similar transposition. For this
the transposeArray2D() function is provided. Here is an example array:

```
[
  [5.1, 4.9, 4.7, 4.6, 5, 5.4],
  [3.5, 3, 3.2, 3.1, 3.6, 3.9],
  [1.4, 1.4, 1.3, 1.5, 1.4, 1.7],
  [0.2, 0.2, 0.2, 0.2, 0.2, 0.4],
  ["setosa", "setosa", "setosa", "setosa", "setosa", "setosa"]
]
```

HTMLWidgets.transposeArray2D() can transpose it to:

```
[
  [5.1, 3.5, 1.4, 0.2, "setosa"],
  [4.9, 3, 1.4, 0.2, "setosa"],
  [4.7, 3.2, 1.3, 0.2, "setosa"],
  [4.6, 3.1, 1.5, 0.2, "setosa"],
  [5, 3.6, 1.4, 0.2, "setosa"],
  [5.4, 3.9, 1.7, 0.4, "setosa"]
]
```

As a real example, the dygraphs[11] widget uses this function to transpose the
"file" (data) argument it gets from the R side before passing it on to the dygraphs library:

---

[11] https://rstudio.github.io/dygraphs

```
renderValue: function(x) {

    // ... code excluded ...

    // transpose array
    x.attrs.file = HTMLWidgets.transposeArray2D(x.attrs.file);

    // ... more code excluded ...
}
```

16.5.1.3 Custom JSON serializer

You may find it necessary to customize the JSON serialization of widget data when the default serializer in **htmlwidgets** does not work in the way you have expected. For widget package authors, there are two levels of customization for the JSON serialization: you can either customize the default values of arguments for jsonlite::toJSON(), or just customize the whole function.

1. jsonlite::toJSON() has a lot of arguments, and we have already changed some of its default values. Below is the JSON serializer we use in **htmlwidgets** at the moment:

```
function(x, ..., dataframe = "columns",
  null = "null", na = "null", auto_unbox = TRUE,
  digits = getOption("shiny.json.digits",
    16), use_signif = TRUE, force = TRUE,
  POSIXt = "ISO8601", UTC = TRUE, rownames = FALSE,
  keep_vec_names = TRUE, strict_atomic = TRUE) {
  if (strict_atomic)
    x <- I(x)
  jsonlite::toJSON(x, dataframe = dataframe,
    null = null, na = na, auto_unbox = auto_unbox,
    digits = digits, use_signif = use_signif,
    force = force, POSIXt = POSIXt,
    UTC = UTC, rownames = rownames,
    keep_vec_names = keep_vec_names,
```

```
    json_verbatim = TRUE, ...)
}
```

For example, we convert data frames to JSON by columns instead of rows (the latter is jsonlite::toJSON's default). If you want to change the default values of any arguments, you can attach an attribute TOJSON_ARGS to the widget data to be passed to createWidget(), e.g.,

```
fooWidget = function(data, name, ...) {
  # ... process the data ...
  params = list(foo = data, bar = TRUE)
  # customize toJSON() argument values
  attr(params, 'TOJSON_ARGS') = list(
    digits = 7, na = 'string'
  )
  htmlwidgets::createWidget(name, x = params, ...)
}
```

We changed the default value of digits from 16 to 7, and na from null to string in the above example. It is up to you, the package author, whether you want to expose such customization to users. For example, you can leave an extra argument in your widget function so that users can customize the behavior of the JSON serializer:

```
fooWidget = function(
  data, name, ..., JSONArgs = list(digits = 7)
) {
  # ... process the data ...
  params = list(foo = data, bar = TRUE)
  # customize toJSON() argument values
  attr(params, 'TOJSON_ARGS') = JSONArgs
  htmlwidgets::createWidget(name, x = params, ...)
}
```

You can also use a global option htmlwidgets.TOJSON_ARGS to customize the JSON serializer arguments for all widgets in the current R session, e.g.

```
options(htmlwidgets.TOJSON_ARGS = list(
  digits = 7, pretty = TRUE
))
```

2. If you do not want to use **jsonlite**, you can completely override the serializer function by attaching an attribute TOJSON_FUNC to the widget data, e.g.,

```
fooWidget = function(data, name, ...) {
  # ... process the data ...
  params = list(foo = data, bar = TRUE)
  # customize the JSON serializer
  attr(params, 'TOJSON_FUNC') = MY_OWN_JSON_FUNCTION
  htmlwidgets::createWidget(name, x = params, ...)
}
```

Here MY_OWN_JSON_FUNCTION can be an arbitrary R function that converts R objects to JSON. If you have also specified the TOJSON_ARGS attribute, it will be passed to your custom JSON function, too.

16.5.2 Passing JavaScript functions

As you would expect, character vectors passed from R to JavaScript are converted to JavaScript strings. However, what if you want to allow users to provide custom JavaScript functions for formatting, drawing, or event handling? For this case, the **htmlwidgets** package includes a JS() function that allows you to request that a character value is evaluated as JavaScript when it is received on the client.

For example, the dygraphs widget (https://rstudio.github.io/dygraphs) includes a dyCallbacks function that allows the user to provide callback functions for a variety of contexts. These callbacks are "marked" as contain-

ing JavaScript so that they can be converted to actual JavaScript functions on
the client:

```
library(dygraphs)
dyCallbacks(
  clickCallback = JS(...)
  drawCallback = JS(...)
  highlightCallback = JS(...)
  pointClickCallback = JS(...)
  underlayCallback = JS(...)
)
```

Another example is in the DT (DataTables) widget (https://rstudio.
github.io/DT), where users can specify an initCallback with JavaScript
to execute after the table is loaded and initialized:

```
datatable(head(iris, 20), options = list(
  initComplete = JS(
    "function(settings, json) {",
    "$(this.api().table().header()).css({",
      'background-color': '#000',
      'color': '#fff'
    });",
    "}")
))
```

If multiple arguments are passed to JS() (as in the above example), they will
be concatenated into a single string separated by \n.

16.5.3   Custom widget HTML

Typically the HTML "housing" for a widget is just a <div> element, and this
is correspondingly the default behavior for new widgets that do not specify
otherwise. However, sometimes you need a different element type. For example, the sparkline widget (https://github.com/htmlwidgets/sparkline)
requires a <span> element, so it implements the following custom HTML generation function:

```
sparkline_html = function(id, style, class, ...){
  tags$span(id = id, class = class)
}
```

Note that this function is looked up within the package implementing the widget by the convention `widgetname_html`, so it need not be formally exported from your package or otherwise registered with **htmlwidgets**.

Most widgets will not need a custom HTML function, but if you need to generate custom HTML for your widget (e.g., you need an `<input>` or a `<span>` rather than a `<div>`), you should use the **htmltools** package (as demonstrated by the code above).

16.5.4 Create a widget without an R package

As we mentioned in Section 16.3, it is possible to create a widget without creating an R package in the first place. Below is an example:

```
#' @param text A character string.
#' @param interval A time interval (in seconds).
blink = function(text, interval = 1) {
  htmlwidgets::createWidget(
    'blink', list(text = text, interval = interval),
    dependencies = htmltools::htmlDependency(
      'blink', '0.1', src = c(href = ''), head = '
<script>
HTMLWidgets.widget({
  name: "blink",
  type: "output",
  factory: function(el, width, height) {
    return {
      renderValue: function(x) {
        setInterval(function() {
          el.innerText = el.innerText == "" ? x.text : "";
        }, x.interval * 1000);
      },
      resize: function(width, height) {}
    };
```

```
  }
});
</script>'
    )
  )
}
```

```
blink('Hello htmlwidgets!', .5)
```

The widget simply shows a blinking character string, and you can specify the time interval. The key of the implementation is the HTML dependency, in which we used the `head` argument to embed the JavaScript binding. The value of the `src` argument is a little hackish due to the current restrictions in **htmltools** (which might be removed in the future). In the `renderValue` method, we show or hide the text periodically using the JavaScript function `setInterval()`.

17

Document Templates

When you create a new R Markdown document from the RStudio menu `File -> New File -> R Markdown`, you will see a default example document (a template) in the RStudio editor. In fact, you can create custom document templates by yourself, which can be useful if you need to create a particular type of document frequently or want to customize the appearance of the final report. The **rticles** package in Chapter 13 is a good example of custom templates for a range of journals. Some additional examples of where a template could be used include:

- Creating a company branded R Markdown template that includes a logo and branding colors.

- Updating the default YAML settings to include standard fields for `title`, `author`, `date`, or default `output` options.

- Customizing the layout of the output document by adding additional fields to the YAML metadata. For example, you can add a `department` field to be included within your title page.

Once created, templates are easily accessed within RStudio, and will appear within the "New R Markdown" window as shown in Figure 17.1.

This chapter explains how to create templates and share them within an R package. If you would like to see some real-world examples, you may check out the source package of **rticles** (`https://github.com/rstudio/rticles`). The `rmarkdown::html_vignette` format is also a relatively simple example (see both its R source code and the template structure[1]). In addition, Michael Harper has kindly prepared more examples in the repository `https://github.com/mikey-harper/example-rmd-templates`.

---

[1]`https://github.com/rstudio/rmarkdown/tree/master/inst/rmarkdown/templates/html_vignette`

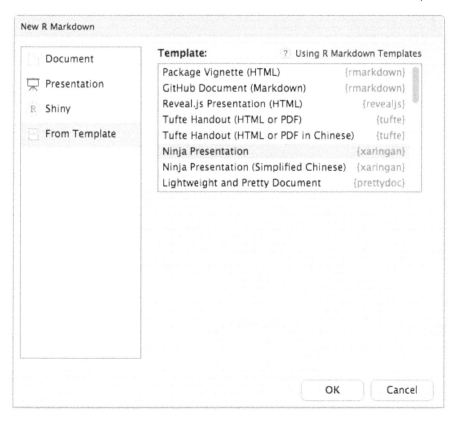

FIGURE 17.1: Selecting R Markdown templates within RStudio.

17.1 Template structure

R Markdown templates should be contained within an R package, which can be easily created from the menu File -> New Project in RStudio (choose the project type to be "R Package"). If you are already familiar with creating R packages, you are certainly free to use your own favorite way to create a new package.

Templates are located within the inst/rmarkdown/templates directory of a package. It is possible to contain multiple templates in a single package, with each template stored in a separate sub-directory. As a minimal example, inst/rmarkdown/templates/my_template requires the following files:

```
template.yaml
skeleton/skeleton.Rmd
```

The `template.yaml` specifies how the template is displayed within the RStudio "From Template" dialog box. This YAML file must have a `name` and a `description` field. You can optionally specify `create_dir: true` if you want to a new directory to be created when the template is selected. As an example of the `template.yaml` file:

```
name: My Template
description: This is my template
```

You can provide a brief example R Markdown document in `skeleton.Rmd`, which will be opened in RStudio when the template is selected. We can add section titles, load commonly used packages, or specify default YAML parameters in this skeleton document. In the following example, we specify the default output format to `bookdown::html_document2`, and select a default template `flatly`:

```
---
title: "Untitled"
author: "Your Name"
output:
  bookdown::html_document2:
    toc: true
    fig_caption: true
    template: flatly
---

## Introduction

## Analysis

## Conclusions
```

17.2 Supporting files

Sometimes a template may require supporting files (e.g., images, CSS files, or LaTeX style files). Such files should be placed in the `skeleton` directory. They will be automatically copied to the directory where the new document is created. For example, if your template requires a logo and CSS style sheet, they can be put under the directory `inst/rmarkdown/templates/my_template`:

```
template.yaml
skeleton/skeleton.Rmd
skeleton/logo.png
skeleton/styles.css
```

We can refer to these files within the `skeleton.Rmd` file, e.g.,

```
---
title: "Untitled"
author: "Your Name"
output:
  html_document:
    css: styles.css
---

![logo](logo.png)

# Introduction

# Analysis

```{r}
knitr::kable(mtcars[1:5, 1:5])
```

Conclusion
```

17.3 Custom Pandoc templates

An R Markdown is first compiled to Markdown through **knitr**, and then converted to an output document (e.g., PDF, HTML, or Word) by Pandoc through a Pandoc template. While the default Pandoc templates used by R Markdown are designed to be flexible by allowing parameters to be specified in the YAML, users may wish to provide their own template for more control over the output format.

You can make use of additional YAML fields from the source document when designing a Pandoc template. For example, you may wish to have a `department` field to be added to your title page, or include an `editor` field to be displayed below the author. We can add additional variables to the Pandoc template by surrounding the variable in dollar signs (`$`) within the template. Most variables take values from the YAML metadata of the R Markdown document (or command-line arguments passed to Pandoc). We may also use conditional statements and for-loops. Readers are recommended to check the Pandoc manual for more details: `https://pandoc.org/MANUAL.html#using-variables-in-templates`. Below is an example of a very minimal Pandoc template for HTML documents that only contains two variables (`$title$` and `$body$`):

```
<html>
  <head>
    <title>$title$</title>
  </head>

  <body>
  $body$
  </body>
</html>
```

For R Markdown to use the customized template, you can specify the `template` option in the output format (provided that the output format supports this option), e.g.,

```
output:
  html_document:
    template: template.html
```

If you wish to design your own template, we recommend starting from the default Pandoc templates included within the **rmarkdown** package (https://github.com/rstudio/rmarkdown/tree/master/inst/rmd) or Pandoc's built-in templates (https://github.com/jgm/pandoc-templates).

17.4 Sharing your templates

As templates are stored within packages, it is easy to distribute them to other users. If you decide not to take the normal approach of publishing your package to CRAN, you may consider using GitHub to host your package instead, in which case users can also easily install your package and templates:

```
if (!requireNamespace("devtools")) install.packages("devtools")
devtools::install_github("username/packagename")
```

To find out more about packages and the use of GitHub, you may refer to the book "*R Packages*" (Wickham, 2015) (http://r-pkgs.had.co.nz/git.html).

18

Creating New Formats

The **rmarkdown** package has included many built-in document and presentation formats. At their core, these formats are just R functions. When you include an output format in the YAML metadata of a document, you are essentially specifying the format function to call and the parameters to pass to it.

We can create new formats for R Markdown, which makes it easy to customize output formats to use specific options or refer to external files. Defining a new function can be particularly beneficial if you have generated a new template as described in Chapter 17, as it allows you to use your custom templates without having to copy any files to your local directory.

18.1 Deriving from built-in formats

The easiest way to create a new format is to write a function that calls one of the built-in formats. These built-in formats are designed to be extensible enough to serve as the foundation of custom formats. The following example, quarterly_report, is based on html_document but alters the default options:

```
quarterly_report = function(toc = TRUE) {
  # locations of resource files in the package
  pkg_resource = function(...) {
    system.file(..., package = "mypackage")
  }

  css    = pkg_resource("reports/styles.css")
  header = pkg_resource("reports/quarterly/header.html")
```

```
# call the base html_document function
rmarkdown::html_document(
  toc = toc, fig_width = 6.5, fig_height = 4,
  theme = NULL, css = css,
  includes = rmarkdown::includes(before_body = header)
)
}
```

The new format defined has the following behavior:

1. Provides an option to determine whether a table of contents should be generated (implemented by passing toc through to the base format).

2. Sets a default height and width for figures (note that this is intentionally not user-customizable so as to encourage a standard for all reports of this type).

3. Disables the default Bootstrap theme and provides custom CSS in its place.

4. Adds a standard header to every document.

Note that (3) and (4) are implemented using external files that are stored within the package that defines the custom format, so their locations need to be looked up using the system.file() function.

18.2 Fully custom formats

Another lower-level approach is to define a format directly by explicitly specifying **knitr** options and Pandoc command-line arguments. At its core, an R Markdown format consists of:

1. A set of **knitr** options that govern how Rmd is converted to Markdown.

2. A set of Pandoc options that govern how Markdown is converted to the final output format (e.g., HTML).

3. Some optional flags and filters (typically used to control handling of supporting files).

You can create a new format using the output_format() function in **rmarkdown**. Here is an example of a simple format definition:

```
#' @importFrom rmarkdown output_format knitr_options pandoc_options
simple_html_format = function() {
  # if you don't use roxygen2 (see above), you need to either
  # library(rmarkdown) or use rmarkdown::
  output_format(
    knitr = knitr_options(opts_chunk = list(dev = 'png')),
    pandoc = pandoc_options(to = "html"),
    clean_supporting = FALSE
  )
}
```

The **knitr** and Pandoc options can get considerably complicated (see help pages ?rmarkdown::knitr_options and ?rmarkdown::pandoc_options for details). The clean_supporting option indicates that you are not creating self-contained output (like a PDF or HTML document with base64 encoded resources), and therefore want to preserve supporting files like R plots generated during knitting.

You can also pass a base_format to the output_format() function if you want to inherit all of the behavior of an existing format but tweak a subset of its options.

If there are supporting files required for your format that cannot be easily handled by the includes option (see Section 3.1.10.2), you will also need to use the other arguments to output_format to ensure they are handled correctly (e.g., use the intermediates_generator to copy them into the place alongside the generated document).

The best way to learn more about creating fully custom formats is to study the source code of the existing built-in formats (e.g., html_document and pdf_document): https://github.com/rstudio/rmarkdown/tree/master/R. In some cases, a custom format will define its own Pandoc template, which was discussed in Section 17.3.

18.3   Using a new format

New formats should be stored within a package and installed onto your local system. This allows the format to be provided to the document YAML. Assuming our example format `quarterly_report` is in a package named **mypackage**, we can use it as follows:

```
---
title: "Habits"
output:
  mypackage::quarterly_report:
    toc: true
---
```

This means to use the `quarterly_report()` function defined in **mypackage** as the output format, and to pass `toc` = `TRUE` as a parameter to the function.

19

Shiny Documents

We have briefly introduced Shiny documents in Section 2.8.2. Shiny is a very powerful framework for building web applications based on R. It is out of the scope of this book to make a comprehensive introduction to Shiny (which is too big a topic). We recommend that readers who are not familiar with Shiny learn more about it from the website `https://shiny.rstudio.com` before reading this chapter.

Unlike the more traditional workflow of creating static reports, you can create documents that allow your readers to change the parameters underlying your analysis and see the results immediately in Shiny R Markdown documents. In the example shown in Figure 2.8, the histogram will be automatically updated to reflect the number of bins selected by the reader.

A picture is worth a thousand words, and a Shiny document can potentially show you a thousand pictures as you interact with it. The readers are no longer tied to the fixed analysis and conclusions in the report. They may explore other possibilities by themselves, and possibly make new discoveries or draw different conclusions.

19.1 Getting started

You can turn any *HTML-oriented* R Markdown documents to Shiny documents by adding `runtime: shiny` to the YAML metadata as a *top-level* field, e.g.,

```
---
title: "Shiny Document"
output: html_document
```

```
runtime: shiny
---
```

Note that the output format of the R Markdown document must be an HTML format. That is, the document should generate a web page (a `*.html` file). Non-HTML formats such as `pdf_document` and `word_document` will not work with the Shiny runtime. Please also note that some presentation formats are also HTML formats, such as `ioslides_presentation`, `slidy_presentation`, and `revealjs::revealjs_presentation`.

You can also create a new Shiny document from the RStudio menu `File ->` `New File -> R Markdown`, and choose the document type "Shiny" (see Figure 19.1).

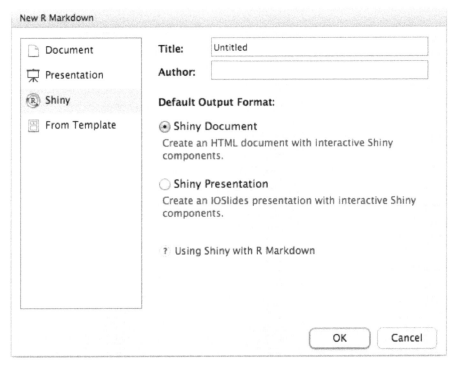

FIGURE 19.1: Create a new Shiny document in RStudio.

To run a Shiny document in RStudio, you need to click the button "Run Document" on the editor toolbar (RStudio will automatically replace the "Knit" button with "Run Document" when it detects a Shiny document). If you do not use RStudio, or want to run the document in the R console for troubleshooting, you can call the function `rmarkdown::run()` and pass the file-name to it.

You can embed Shiny inputs and outputs in your document. Outputs are automatically updated whenever inputs change. In the following example, we create a numeric input (`numericInput`) with the name rows, and then refer to its value via `input$rows` when generating output:

```r
```{r, echo=FALSE}
numericInput("rows", "How many cars?", 5)

renderTable({
  head(cars, input$rows)
})
```
```

How many cars?

7

mpg	cyl	disp	hp	drat	wt	qsec	vs	am	gear	carb
21.00	6.00	160.00	110.00	3.90	2.62	16.46	0.00	1.00	4.00	4.00
21.00	6.00	160.00	110.00	3.90	2.88	17.02	0.00	1.00	4.00	4.00
22.80	4.00	108.00	93.00	3.85	2.32	18.61	1.00	1.00	4.00	1.00
21.40	6.00	258.00	110.00	3.08	3.21	19.44	1.00	0.00	3.00	1.00
18.70	8.00	360.00	175.00	3.15	3.44	17.02	0.00	0.00	3.00	2.00
18.10	6.00	225.00	105.00	2.76	3.46	20.22	1.00	0.00	3.00	1.00
14.30	8.00	360.00	245.00	3.21	3.57	15.84	0.00	0.00	3.00	4.00

FIGURE 19.2: Increase the number of rows in the table in a Shiny document.

In the above example, the output code was wrapped in a call to `renderTable()`. There are many other render functions in Shiny that can be used for plots, printed R output, and more. This example uses `renderPlot()` to create dynamic plot output:

```r
```{r, echo=FALSE}
sliderInput("bins", "Number of bins:", 30, min = 1, max = 50)
```
```

```
renderPlot({
  x    = faithful[, 2]   # Old Faithful Geyser data
  bins = seq(min(x), max(x), length.out = input$bins + 1)

  # draw the histogram with the specified number of bins
  hist(x, breaks = bins, col = 'darkgray', border = 'white')
})
```
```

**Number of bins:**

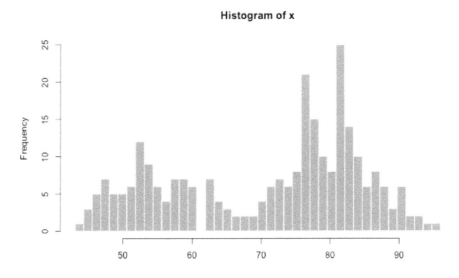

**FIGURE 19.3:** Change the number of bins of a histogram in a Shiny document.
```

19.2   Deployment

Shiny documents must be supported by a live R session running behind the scenes. When you run a Shiny document locally, it uses your local R session. Usually only you can see and interact with the document. If you want to share the document with people who do not have R installed, or do not want to run the document locally, you will have to deploy the document on a server, and share the URL of the document. Then other people only need a web browser to visit your document.

There are two ways to deploy a Shiny document. You can either use a hosted service provided by RStudio, or set up your own server. The first way is technically easier, but sometimes you may not be allowed to use an external hosted service, so you have to install the required software (Shiny Server or RStudio Connect) on your own server to deploy the Shiny documents.

19.2.1   ShinyApps.io

You can publish Shiny documents to the ShinyApps (`https://shinyapps.io`) hosted service. To do this you, should ensure that you have:

1. An account on ShinyApps (use the signup form to request an account).

2. A latest version of the **rsconnect** R package. You can install this as follows:

    ```r
    install.packages("rsconnect")
    ```

You can then deploy an interactive Shiny document in the same way that you currently deploy normal Shiny applications. From the working directory containing the document(s), just execute:

```r
rsconnect::deployApp()
```

If you are using RStudio, you can also use the `Publish` button available at the top-right of the window when running a Shiny document (see Figure 19.4).

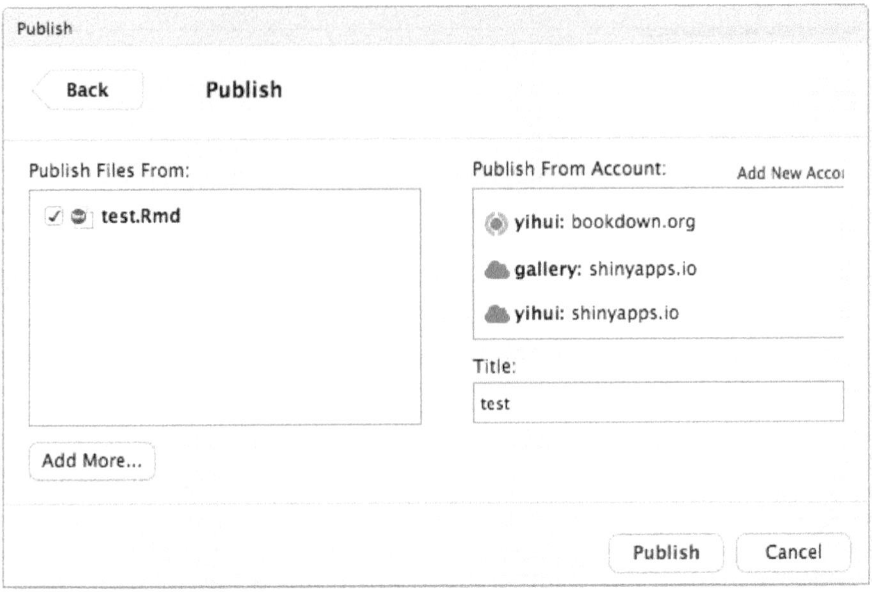

FIGURE 19.4: Deploy a Shiny document to ShinyApps.io.

If there is a file named index.Rmd in the directory, it will be served as the default document for that directory, otherwise an explicit path to the Rmd file should be specified in the URL if you want to visit this Rmd document. For example, the URL for index.Rmd deployed to ShinyApps may be of the form https://example.shinyapps.io/appName/, and the URL for test.Rmd may be of the form https://example.shinyapps.io/appName/test.Rmd.

19.2.2   Shiny Server / RStudio Connect

Both Shiny Server (https://www.rstudio.com/products/shiny/shiny-server/) and RStudio Connect (https://www.rstudio.com/products/connect/) can be used to publish Shiny documents. They require knowledge about Linux. Installing and configuring them should normally be a task for your system administrator if you are not familiar with Linux or do not have the privilege.

19.3   Embedded Shiny apps

Besides embedding individual Shiny inputs and outputs in R Markdown, it is also possible to embed a standalone Shiny application within a document. There are two ways to do this:

1. Defining the application inline using the `shinyApp()` function; or

2. Referring to an external application directory using the `shinyAppDir()` function.

Both functions are available in the **shiny** package (not **rmarkdown**), which will be automatically loaded when `runtime: shiny` is specified in the YAML metadata of the document, so you do not have to call `library(shiny)` to load **shiny** (although it does not hurt if you load a package twice).

19.3.1   Inline applications

This example uses an inline definition:

````
```{r, echo=FALSE}
shinyApp(

  ui = fluidPage(
    selectInput("region", "Region:",
                choices = colnames(WorldPhones)),
    plotOutput("phonePlot")
  ),

  server = function(input, output) {
    output$phonePlot = renderPlot({
      barplot(WorldPhones[,input$region]*1000,
              ylab = "Number of Telephones", xlab = "Year")
    })
  },

  options = list(height = 500)
````
````

```
)
```

Note the use of the `height` parameter to determine how much vertical space the embedded application should occupy.

19.3.2   External applications

This example embeds a Shiny application defined in another directory:

```
```{r, echo = FALSE}
shinyAppDir(
  system.file("examples/06_tabsets", package="shiny"),
  options = list(width = "100%", height = 700)
)
```
```

Note that in all of R code chunks above, the chunk option `echo = FALSE` is used. This is to prevent the R code within the chunk from rendering to the output document alongside the Shiny components.

19.4   Shiny widgets

Shiny widgets enable you to create re-usable Shiny components that are included within an R Markdown document using a single function call. Shiny widgets can also be invoked directly from the console (useful during authoring) and show their output within the RStudio Viewer pane or an external web browser.

19.4.1   The `shinyApp()` function

At their core, Shiny widgets are mini-applications created using the `shinyApp()` function. Rather than creating a `ui.R` and `server.R` (or `app.R`) as you would for a typical Shiny application, you pass the UI and server definitions to the `shinyApp()` function as arguments. We have given an example in Section 19.3.1.

The simplest type of Shiny widget is just an R function that returns a `shinyApp()`.

19.4.2 Example: k-Means clustering

The **rmdexamples** package (`https://github.com/rstudio/rmdexamples`) includes an example of a Shiny widget implemented in this fashion. The `kmeans_cluster()` function takes a single `dataset` argument and returns a Shiny widget to show the result of k-Means clustering. You can use it within an R Markdown document like this:

```
```{r, echo = FALSE}
library(rmdexamples)
kmeans_cluster(iris)
```
```

Figure 19.5 shows what the widget looks like inside a running document.

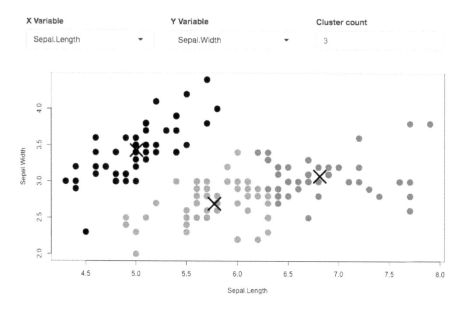

FIGURE 19.5: A Shiny widget to apply k-Means clustering on a dataset.

Below is the source code of the kmeans_cluster() function:

```r
kmeans_cluster = function(dataset) {

  library(shiny)
  vars = names(dataset)

  shinyApp(
    ui = fluidPage(
      fluidRow(style = "padding-bottom: 20px;",
        column(4, selectInput('xcol', 'X Variable', vars)),
        column(4, selectInput('ycol', 'Y Variable', vars,
                              selected = vars[2])),
        column(4, numericInput('clusters', 'Cluster count', 3,
                              min = 1, max = 9))
      ),
      fluidRow(
        plotOutput('kmeans', height = "400px")
      )
    ),

    server = function(input, output, session) {

      # Combine the selected variables into a new data frame
      selectedData = reactive({
        dataset[, c(input$xcol, input$ycol)]
      })

      clusters = reactive({
        kmeans(selectedData(), input$clusters)
      })

      output$kmeans = renderPlot(height = 400, {
        res = clusters()
        par(mar = c(5.1, 4.1, 0, 1))
        plot(selectedData(),
             col = res$cluster, pch = 20, cex = 3)
        points(res$centers, pch = 4, cex = 4, lwd = 4)
      })
```

```
  },

  options = list(height = 500)
  )
}
```

19.4.3   Widget size and layout

Shiny widgets may be embedded in various places including standard full width pages, smaller columns within pages, and even HTML5 presentations. For the widget size and layout to work well in all of these contexts, we recommend that the total height of the widget is no larger than 500 pixels. This is not a hard and fast rule, but HTML5 slides can typically only display content less than 500px in height, so if you want your widget to be usable within presentations, this is a good guideline to follow.

You can also add an explicit height argument to the function that creates the widget (default to 500 so it works well within slides).

19.5   Multiple pages

You can link to other Shiny documents by using the Markdown link syntax and specifying the *relative* path to the document, e.g., [Another Shiny Document](another.Rmd). If you click the link to another Rmd document on one page, that Rmd document will be launched as the current interactive Shiny document.

Currently, only one document can be active at a time, so documents cannot easily share state, although some primitive global sharing is possible via the R script global.R (see the help page ?rmarkdown::run).

By default, it is only possible to link to R Markdown files in the same directory subtree as the file on which rmarkdown::run() was invoked (e.g., you cannot link to ../foo.Rmd). You can use the dir argument of rmarkdown::run() to indicate the directory to be treated as the root.

19.6   Delayed rendering

A Shiny document is typically rendered every time it is shown, and is not shown to the user until the rendering is complete. Consequently, a document that is large or contains expensive computations may take some time to load.

If your document contains interactive Shiny components that do not need to be rendered right away, you can wrap Shiny code in the `rmark-down::render_delayed()` function. This function saves its argument until the document's rendering is done and has been shown to the user, then evaluates it and injects it into the output document when the computation is finished.

Here is an example that demonstrates how `render_delayed()` works. The code enclosed within the `render_delayed()` call will execute only after the document has been loaded and displayed to the user:

````
```{r, echo = FALSE}
rmarkdown::render_delayed({
  numericInput("rows", "How many cars?", 5)

  renderTable({
    head(cars, input$rows)
  })
})
```
````

19.7   Output arguments for render functions

In a typical Shiny application, you specify an output element in the UI using functions like `plotOutput()` and `verbatimTextOutput()`, and render its content using functions like `renderPlot()` and `renderPrint()`.

By comparison, in a Shiny document, the UI elements are often implicitly and automatically created when you call the `renderXXX()` functions. For example, you may want to use a renderPlot() function without having to create a plotOutput() slot beforehand. In this case, Shiny helpfully associates the corresponding output object to each renderXXX() function, letting you use Shiny code outside of a full Shiny app. However, some functionality can be lost in this process. In particular, plotOutput() can take in some optional arguments to set things like width and height, or allow you to click or brush over the plot (and store that information).

To pass options from renderXXX() to xxxOutput(), you can use the outputArgs argument, if it is available to specific renderXXX() functions. For example, suppose that you want to render a plot and specify its width to be 200px and height to be 100px. Then you should use:

```
```{r, echo = FALSE}
renderPlot({
  plot(yourData)
}, outputArgs = list(width = "200px", height = "100px")
)
```
```

No matter how many output arguments you want to set (all the way from zero to all possible ones), outputArgs always takes in a list (the default is an empty list, which sets no output arguments). If you try to pass in a non-existent argument, you will get an error like the following message (in this example, you tried to set an argument named not_an_argument):

```
**Error**: Unused argument: in `outputArgs`, `not_an_argument`
is not an valid argument for the output function
```

To see outputArgs in action, run the R Markdown document below or visit https://gallery.shinyapps.io/output-args/ for the live version online. The document is interactive: brush over the image and see the xmin, xmax, ymin, and ymax values change (printed right under the image).

```
---
title: Setting output args via render functions
runtime: shiny
output: html_document
---
```

This interactive Rmd document makes use of the `outputArgs` argument now available to all Shiny `render` functions. To give an example, this allows you to set arguments to `imageOutput` through `renderImage`. This means that you don't have to create a `ui` object just to be able to brush over an image. Note that this only applies to snippets of Shiny code during an interactive Rmd (and not to embedded full apps -- the ones you need to call `shinyApp` to run).

Brushing over an image (and storing the data)

````
```{r setup, echo=FALSE}
library(datasets)

generateImage = function() {
  outfile = tempfile(fileext = '.png')
  png(outfile)
  par(mar = c(0,0,0,0))
  image(volcano, axes = FALSE)
  contour(volcano, add = TRUE)
  dev.off()
  list(src = outfile)
}
```
````

````
```{r image}
renderImage({
  generateImage()
}, deleteFile = TRUE,
   outputArgs = list(brush = brushOpts(id = "plot_brush"),
                     width = "250",
                     height = "250px")
)
```
````

Here is some of the brushing info sent to the server:
(brush over the image to change the data)

````
```{r brush info}
renderText({
  print(input$plot_brush)
  brush = input$plot_brush
  paste0("xmin: ", brush$xmin, "; ",
         "xmax: ", brush$xmax, "; ",
         "ymin: ", brush$ymin, "; ",
         "ymax: ", brush$ymax)
})
```
````

Resizing a plot

````
```{r plot}
renderPlot({
  plot(cars)
}, outputArgs = list(width = "75%",
                     height = "250px")
)
```
````

19.7.1 A caveat

We want to emphasize that you can only use this functionality within a Shiny
R Markdown document (i.e., you must set `runtime: shiny` in the YAML
metadata). But even if that is the case, this is only applicable to pieces of Shiny
code that render output without the corresponding explicit output elements
in the UI. If you embed a full Shiny application in your document and try to
use `outputArgs`, it will be ignored and print the following warning to the R
Markdown console (in this case, your `ui` function would be something like
`ui = plotOutput("plot")`):

```
Warning in `output$plot`(...) :
Unused argument: outputArgs. The argument outputArgs is only
meant to be used when embedding snippets of Shiny code in an
```

R Markdown code chunk (using runtime: shiny). When running a
full Shiny app, please set the output arguments directly in
the corresponding output function of your UI code.

The same will happen if you try to use outputArgs in any other context, such
as inside a regular (i.e., not embedded) Shiny app. The rationale is that if you
are already specifying a ui function with all the output objects made explicit,
you should set their arguments directly there instead of going through this
round-about way.

Bibliography

Allaire, J., Horner, J., Marti, V., and Porte, N. (2017). *markdown: 'Markdown' Rendering for R*. R package version 0.8.

Allaire, J., R Foundation, Wickham, H., Journal of Statistical Software, Xie, Y., Vaidyanathan, R., Association for Computing Machinery, Boettiger, C., Elsevier, Broman, K., Mueller, K., Quast, B., Pruim, R., Marwick, B., Wickham, C., Keyes, O., Yu, M., Emaasit, D., Onkelinx, T., Desautels, M.-A., Leutnant, D., and MDPI (2018a). *rticles: Article Formats for R Markdown*. R package version 0.4.2.9000.

Allaire, J., Ushey, K., and Tang, Y. (2018b). *reticulate: Interface to 'Python'*. R package version 1.8.

Allaire, J., Xie, Y., McPherson, J., Luraschi, J., Ushey, K., Atkins, A., Wickham, H., Cheng, J., Chang, W., and Iannone, R. (2018c). *rmarkdown: Dynamic Documents for R*. https://rmarkdown.rstudio.com, https://github.com/rstudio/rmarkdown.

Barnier, J. (2017). *rmdformats: HTML Output Formats and Templates for 'rmarkdown' Documents*. R package version 0.3.3.

Bion, R., Chang, R., and Goodman, J. (2018). How r helps airbnb make the most of its data. *The American Statistician*, 72(1):46–52.

Borges, B. and Allaire, J. (2017). *flexdashboard: R Markdown Format for Flexible Dashboards*. R package version 0.5.1.

Borges, B. and Allaire, J. (2018). *learnr: Interactive Tutorials for R*. R package version 0.9.2.

Chang, W. (2017). *webshot: Take Screenshots of Web Pages*. R package version 0.5.0.

Chang, W., Cheng, J., Allaire, J., Xie, Y., and McPherson, J. (2018). *shiny: Web Application Framework for R*. R package version 1.1.0.

Cheng, J., Karambelkar, B., and Xie, Y. (2018). *leaflet: Create Interactive Web Maps with the JavaScript 'Leaflet' Library*. R package version 2.0.1.

El Hattab, H. and Allaire, J. (2017). *revealjs: R Markdown Format for 'reveal.js' Presentations*. R package version 0.9.

Hartgerink, C. H., Wicherts, J. M., and van Assen, M. A. (2017). Too good to be false: Nonsignificant results revisited. *Collabra: Psychology*, 3(1).

Knuth, D. E. (1984). Literate programming. *The Computer Journal*, 27(2):97–111.

Li, C. (2018). *JuliaCall: Seamless Integration Between R and 'Julia'*. R package version 0.13.0.

Lowndes, J. S. S., Best, B. D., Scarborough, C., Afflerbach, J. C., Frazier, M. R., O'Hara, C. C., Jiang, N., and Halpern, B. S. (2017). Our path to better science in less time using open data science tools. *Nature ecology & evolution*, 1(6).

Qiu, Y. (2018). *prettydoc: Creating Pretty Documents from R Markdown*. R package version 0.2.1.

R Core Team (2018). *R: A Language and Environment for Statistical Computing*. R Foundation for Statistical Computing, Vienna, Austria.

RStudio and Inc. (2017). *htmltools: Tools for HTML*. R package version 0.3.6.

Vaidyanathan, R., Xie, Y., Allaire, J., Cheng, J., and Russell, K. (2018). *htmlwidgets: HTML Widgets for R*. R package version 1.2.1.

Vanderkam, D., Allaire, J., Owen, J., Gromer, D., Shevtsov, P., and Thieurmel, B. (2017). *dygraphs: Interface to 'Dygraphs' Interactive Time Series Charting Library*. R package version 1.1.1.4.

Wickham, H. (2015). *R Packages*. O'Reilly Media, Inc., 1st edition.

Wickham, H. and Hesselberth, J. (2018). *pkgdown: Make Static HTML Documentation for a Package*. R package version 1.1.0.

Xie, Y. (2015). *Dynamic Documents with R and knitr*. Chapman and Hall/CRC, Boca Raton, Florida, 2nd edition. ISBN 978-1498716963.

Xie, Y. (2016). *bookdown: Authoring Books and Technical Documents with R Markdown*. Chapman and Hall/CRC, Boca Raton, Florida. ISBN 978-1138700109.

Xie, Y. (2018a). *blogdown: Create Blogs and Websites with R Markdown*. R package version 0.6.12.

Xie, Y. (2018b). *bookdown: Authoring Books and Technical Documents with R Markdown*. R package version 0.7.11.

Xie, Y. (2018c). *DT: A Wrapper of the JavaScript Library 'DataTables'*. R package version 0.4.14.

Xie, Y. (2018d). *knitr: A General-Purpose Package for Dynamic Report Generation in R*. R package version 1.20.5.

Xie, Y. (2018e). *servr: A Simple HTTP Server to Serve Static Files or Dynamic Documents*. R package version 0.10.

Xie, Y. (2018f). *tinytex: Helper Functions to Install and Maintain TeX Live, and Compile LaTeX Documents*. R package version 0.5.8.

Xie, Y. (2018g). *xaringan: Presentation Ninja*. R package version 0.6.7.

Xie, Y. and Allaire, J. (2018). *tufte: Tufte's Styles for R Markdown Documents*. R package version 0.3.

Xie, Y., Hill, A. P., and Thomas, A. (2017). *blogdown: Creating Websites with R Markdown*. Chapman and Hall/CRC, Boca Raton, Florida. ISBN 978-0815363729.

Index